VIDEO
BOOK

Tony Northrup's Adobe Photoshop
Lightroom 5
Training for Photographers

Published by:

Mason Press, Inc.
139 Oswegatchie Rd.
Waterford, CT 06385

ISBN: 978-0-9882634-8-2

Printed and bound in the United States of America by Signature Book Printing, www.sbpbooks.com

Editors: Tanya Egan Gibson, Kyle Guynn
Designer: Chelsea Northrup
Producer: Justin Eckert

Trademarks

For Justin Eckert. You're awesome.

INTRODUCTION   IX

ACKNOWLEDGEMENTS   IX

1 QUICK START

BUYING LIGHTROOM   1

IMPORTING EXISTING PICTURES   1

IMPORTING PICTURES FROM A CAMERA   3

WHERE DID MY MENUS GO? (SHIFT+F)   3

WHERE DID MY PANELS GO?   3

WHERE DID MY PICTURES GO?   4

NAVIGATING LIGHTROOM   4

THE QUICK DEVELOP PANEL   7

UNDOING MISTAKES   8

FIXING MANY PICTURES AT ONCE   8

FINDING PICTURES   10

KEYWORDING PICTURES   11

DEVELOPING PICTURES   11

EXPORTING AND SHARING PICTURES   14

SAVING DISK SPACE   15

2 BASIC ADJUSTMENTS

USING THE HISTOGRAM PANEL   17

USING THE QUICK DEVELOP AND BASIC PANELS   20

USING VIRTUAL COPIES   19

USING STACKS   20

3 ORGANIZING YOUR PICTURES

KEYWORD SETS   33

USING THE KEYWORD LIST PANEL   34

USING THE METADATA PANEL   35

RATINGS   36

COLOR LABELS   37

FLAGS   38

USING THE FILTER BAR   38

PAINTING   49

4 COLLECTIONS & FOLDERS

CREATING COLLECTIONS AND COLLECTION SETS   41

FOLDERS   48

5   CROPPING, ROTATING, & REMOVING SPOTS

    CROPPING AND ROTATING 51

    REMOVING SPOTS 58

    ADJUSTING FEATHERING 59

    ADJUSTING OPACITY 60

6   FILTERS & ADJUSTMENT BRUSHES

    USING GRADUATED FILTERS 63

    USING RADIAL FILTERS 65

    USING THE ADJUSTMENT BRUSH 66

7   THE TONE CURVE PANEL

    ADJUSTING THE TONE CURVE 77

    ADJUSTING THE CONTRAST 80

    ADJUSTING THE SPLITS 80

    CREATING A CUSTOM POINT CURVE 81

    DIRECTLY ADJUSTING PARTS OF YOUR PICTURE 83

    ADJUSTING INDIVIDUAL CHANNELS 84

8   COLOR, SPLIT TONING, & EFFECT

    ADJUSTING LUMINANCE 89

    ADJUSTING SATURATION 90

    LUMINANCE, SATURATION, AND VISUAL WEIGHT 91

    SPOT COLOR/SELECTIVE COLOR 92

    ADJUSTING HUE 93

    RESETTING CHANGES 94

    USING THE ALL AND COLOR TABS 94

    USING SPLIT TONING 95

    ADDING VIGNETTING 96

    ADDING GRAIN 100

9   BLACK & WHITE

    CREATING BLACK-AND-WHITE LANDSCAPES 103

    CREATING BLACK-AND-WHITE WEDDING PHOTOGRAPHS 107

    CREATING BLACK-AND-WHITE STREET PHOTOGRAPHY 108

    CREATING BLACK-AND-WHITE PORTRAITS 110

10 SHARPENING & NOISE REDUCTION

Sharpening 115

Sharpening Radius 117

Sharpening Detail 117

Sharpening Masking 118

Noise Reduction 119

11 LENS CORRECTION & CAMERA CALIBRATION

Lens Corrections 125

Camera Calibration 134

12 PRESETS, HISTORY, SNAPSHOTS, AND BEFORE & AFTER

Before & After 137

History 139

Snapshots 139

Presets 140

13 CUSTOMIZING LIGHTROOM

Changing the Identity Plate 145

Changing the Module Headers 146

Hiding Panels 147

Replace the End Marks 149

Customizing the Toolbars 150

Changing Library and Develop View Options 150

14 CHANGING PREFERENCES

General 159

Presets 160

External Editing 162

File Handling 164

Interface 166

Lightroom Mobile 169

15 IMPORTING, EXPORTING, AND WATERMARKING PICTURES

Importing from a Memory Card 171

Importing from another Catalog 173

Exporting Files 173

Export Location 175

Editing Pictures in Photoshop 181

Publishing Pictures 183

16   TETHERING

TETHERING WITH USB   187

AUTO-IMPORT WITH WI-FI   187

17   PLAYING AND EDITING VIDEOS

PLAYING VIDEOS   193

SETTING THE THUMBNAIL FOR A VIDEO   194

EXTRACTING A STILL PHOTO FROM VIDEO   195

ADJUSTING THE COLOR AND BRIGHTNESS OF VIDEO   195

TRIMMING THE ENDS OF VIDEO   196

ADDING CUTS AND MUSIC TO VIDEO   197

SHARING YOUR VIDEO   198

18   THE MAP MODULE

FINDING TAGGED PICTURES   202

MANUALLY TAGGING LOCATIONS   202

SYNCING LOCATION DATA FROM YOUR SMARTPHONE   202

19   MAKING PHOTO BOOKS

CHOOSING A PRINTING SERVICE   206

CHOOSING A BOOK SIZE   207

CHOOSING PRINTING OPTIONS   207

USING AUTO LAYOUT   208

ADJUSTING PICTURES   209

CHANGING PAGE LAYOUTS   209

ADDING PAGES AND PAGE NUMBERS   209

ADDING TEXT   210

SETTING THE BACKGROUND   210

ADDRESSING RESOLUTION PROBLEMS   211

PRINTING YOUR BOOK   211

20   SLIDESHOWS AND TIME LAPSES

THE IMPROMPTU SLIDESHOW (THAT I DON'T RECOMMEND)   213

CREATING YOUR SIDESHOW   214

CHANGING THE ORDER OF YOUR PICTURES   216

THE OPTIONS PANEL   216

THE LAYOUT PANEL   216

THE OVERLAYS PANEL   217

THE BACKDROP PANEL   217

THE TITLES PANEL 219

THE PLAYBACK PANEL 219

ADDING TEXT 220

MAKING A VIDEO 220

MAKING A TIME LAPSE 221

21   PRINTING PHOTOS AND CREATING CUSTOM LAYOUTS

PRINTING SINGLE PICTURES 223

CHOOSING YOUR PAPER, SETTING UP YOUR PRINTER, AND PRINTING . . . 224

PRINTING CUSTOM-SIZED PICTURES 226

PRINTING MULTIPLE PICTURES ON A PAGE 226

PRINTING PICTURE PACKAGES 227

WATERMARKING YOUR PRINTS 228

MAKING CUSTOM LAYOUTS 229

SOFT-PROOFING 232

22   THE WEB MODULE

WHY YOU'LL HATE IT 225

23   BACKUPS, CATALOGS, & PREVIEWS

BACKING UP YOUR PICTURES 237

MANAGING PREVIEWS 238

USING CATALOGS 241

24   TIPS & TRICKS

DRAG NUMBERS 245

DIRECTLY ADJUST PARTS OF YOUR PICTURES 245

RESET SETTINGS 245

APPLY SHARPENING TO ONLY THE DETAILED PARTS OF YOUR PICTURE . . 246

MAXIMIZE SCREEN SPACE 246

MATCHING EXPOSURES 247

SEE CLIPPED WHITES AND BLACKS 247

DRAG THE HISTOGRAM 248

DRAG PHOTOS INTO LIGHTROOM 249

USE THE SPACE BAR TO DRAG YOUR PHOTOS 249

QUICKLY CHANGE THE DEFAULT SETTINGS FOR NEW PICTURES 249

SAVE 15% OF YOUR DISK SPACE WITH DNGS 249

USE RAW FILES FROM YOUR BRAND NEW CAMERA 251

GET THAT JPG LOOK 251

To get the videos, ebook, & presets, scan the QR code or open this link:

SDP.io/LR5Intro

Thanks for choosing our book! We hope we help you get the most out of your photography and Lightroom.

You already have full access to the videos; just use the links at the top of this page and throughout this book. We need to verify your proof-of-purchase to give you access to the ebook, presets, and Facebook group:

1. If you haven't yet, create an account at SDPCommunity.com by visiting *SDP.io/Register*.

2. Request to join the private Facebook group at *SDP.io/LRFB* (optional).

3. Send us an email at tony@northrup.org with these three items:

   - A receipt or a snapshot of you with this book.
   - The email address you used to register in step 1.
   - Your full name as it appears on Facebook (if you did step 2).

Depending on your learning style, you can use this video book in different ways:

- Watch the 12+ hours of video training at *SDP.io/LR5videos*, and use the book as quick reference.
- Read the book and watch specific videos when you want to better understand a topic.
- Skip to the Index at the back of this book and look up specific topics that you want to learn more about.

Throughout this book, you'll see links to the book's videos and online content. Type the URL into your browser or scan the QR code with your smartphone or tablet. If you haven't used QR codes before, they're just an easier way to type a link to a website. Find a free app by searching your app store for "QR."

Here are some ways to stay in touch with us:

- Subscribe to our YouTube channel at *http://sdp.io/yt* to see our live weekly show where we edit your pictures in Lightroom.
- Like our Facebook page at *http://fb.com/NorthrupPhotography*.
- Follow us on Twitter at @TonyNorthrup and @ChelseaNorthru (there's no 'p' in her last name).

If you find this book useful, please tell your friends. If you love this book, a 5-star review really helps. If you didn't love it, please tell me (tony@northrup.org) and I'll do my best to fix it.

ACKNOWLEDGEMENTS

Many readers gave feedback that helped us improve the quality of this book, including Kyle Guynn, Logan Cartwright, Lec Cel, Bryan Buttigieg, Stuart Patterson, Bonnie Bradley, Charles Clark, Daniel Lorch, Joanna Malukiewicz, Wladimir Paripski, and Kyle Guynn. Yes, I thanked Kyle Guynn twice, because he deserves it.

Creating this has been a massive effort requiring the skills of an expert team. In no particular order:

- Chelsea Northrup, co-photographer, designer, and muse
- Justin Eckert, lead video producer and buddy
- Siobhan Midgett, customer support, video editor, and close friend
- Richard Sullivan, video editor
- Doug Lively, video editor
- Tanya Gibson, editor and proofreader
- Phil Nanzetta (at Signature Book Printing), our printer

1
Lightroom Quick Start

To view the videos that accompany this chapter, scan the QR code or follow the link:

SDP.io/LR5Ch1

This chapter and the accompanying video training will quickly teach you the basics of Adobe Lightroom 5. First, we'll cover the basics of buying Lightroom. Then I'll describe each of the different modules at a high level, so you can decide to jump directly to the chapters of the book that cover the parts of Lightroom most important to your specific needs.

As with the rest of the book, I encourage you to skip directly to the parts you're most interested in. If you prefer learning by watching videos, you can watch the Lesson 1 video and entirely skip this chapter.

Here's a bit of Lightroom trivia: the name is a play on the word "darkroom," the completely dark closet or basement that film photographers use for developing and printing their film. With Lightroom, you no longer have to be in the dark when sorting, cropping, and printing your pictures.

BUYING LIGHTROOM

This section describes the different options for buying Lightroom. If you're currently using the free trial, it will expire after 30 days and then refuse to open. (You can still access your pictures directly, however). Understanding the purchase options can save you a substantial amount of money and headache.

DOWNLOAD VS. DISK

Adobe distributes Lightroom both across the Internet and using DVDs. Some people who purchase Lightroom feel more comfortable buying a DVD because they have something physical that represents their purchase. However, simply owning the physical DVD doesn't allow you to use the software—you need either a license key or an Adobe account with an active license.

You'll always be able to download Lightroom from the Internet. As long as you have Internet access, the download option is faster, more efficient, and more reliable than buying physical media.

PURCHASE VS. LEASE

If you're comfortable downloading software, you have two options for paying Adobe for Lightroom:

- **Purchase**. You can pay a one-time fee (about $75 for teachers and students at *sdp.io/lr5e*, or $145 for everyone else at *sdp.io/lr5*) to buy a permanent license to use Lightroom 5 on one desktop and one mobile computer. If Adobe releases a Lightroom 6, you wouldn't get it for free, but you could continue to use Lightroom 5. If you want the upgrade, you would need to buy it. Upgrade prices are typically about half the purchase price.

- **Lease**. You can pay a monthly or annual fee ($9.99 per month in the US with an annual contract, more elsewhere) for the Creative Cloud Photography plan (*sdp.io/adobedeal*) to use the current version of Lightroom and Photoshop. Photoshop works alongside Lightroom to provide more serious photo editing. You get all updates to both apps for free. You also get access to Lightroom Mobile, which can synchronize some photos to a smartphone or tablet. Don't lease just to get access to Lightroom Mobile. At the moment, it's not very good.

The lease option is less expensive if you need both Lightroom and Photoshop and you plan to keep your software up to date. If you don't plan to use Photoshop, and you don't care about upgrading to new versions, the purchase option is less expensive.

One note about the updates: you might need the latest version of Lightroom to work with a new camera. In other words, even if you never plan to update Lightroom, if you buy a new camera next year, you might need to upgrade Lightroom to read the camera's raw files.

IMPORTING EXISTING PICTURES

The first time you use Lightroom, it will be empty. Lightroom doesn't automatically find pictures on your computer. Before you can browse or edit your pictures, you must import them.

From the **File** menu, select **Import Photos And Video** (**Ctrl+Shift+I** on a PC or **Cmd+Shift+I** on a Mac) to open the Import dialog. Now, expand the Source panel on the left side of the window and select the folder containing your pictures, as shown in the following figure.

If you don't see the folder you've saved your pictures in, you can browse for it using the Source panel on the left side of the Import dialog. On most Windows PCs, your pictures are automatically saved in C:\Users\<*username*>\ My Pictures, as shown in the following figure.

Once you've selected the folder with your pictures, be sure to select the **Include Subfolders** checkbox at the top of the Source panel. Otherwise, you might not see all your pictures.

There are many options on the Import dialog box, and we'll cover them all in detail in Chapter 15. For the meantime, you can simply click Import in the lower-right corner, and Lightroom will start sorting through your pictures. This process will probably take at least a few minutes, but if you have thousands of pictures, it might take hours.

If you have pictures in other folders, don't worry; after you import your first folder, you can repeat this process to import other folders on your computer.

IMPORTING PICTURES FROM A CAMERA

If you have pictures stored on a camera, you can use Lightroom to copy them to your computer and automatically import them into the Lightroom library. You no longer need to use the Windows or Mac OS tools for importing your pictures.

You can connect your camera to your computer in one of two ways:

- **Memory card reader**. Removing your memory card from your camera and inserting it into a memory card reader on your computer is usually the fastest and easiest way to copy pictures from your camera. Many new computers have a memory card reader built in. If yours doesn't have a reader that matches your memory card, just buy a memory card reader from an electronics store. I recommend buying a USB 3 memory card reader (such as *sdp.io/mcr*).

- **USB cable**. Most cameras come with a USB cable to attach the camera directly to your computer. If you've lost the USB cable, you can buy another matching cable from an electronics store. Usually, you just need to connect your camera to your computer and turn the camera on.

You can also use tethering to connect your camera using a wired or wireless network connection, but that's not as fast or convenient. We'll discuss tethering later in this book.

WHERE DID MY MENUS GO? (SHIFT+F)

The single most common question I get about Lightroom is, "Where did my menus go???" No matter which module you're using, press Shift+F repeatedly to cycle through the three different viewing modes, which allow you to hide the title bar and menus, leaving more room for your pictures.

WHERE DID MY PANELS GO?

Lightroom will make you want to get a bigger monitor; you'll always wish you could see more of your pictures. A cheaper alternative is to hide panels that you don't need by clicking the triangles at the top, bottom, and sides of Lightroom, as shown below. To show them again, click the arrow.

WHERE DID MY PICTURES GO?

If all your pictures seem to disappear, you probably need to select **All Photographs** in the Catalog panel, as shown next. By default, when you import new pictures, Lightroom selects the Previous Import catalog. If you want to see your old pictures again, you must select All Photographs.

NAVIGATING LIGHTROOM

Now that you have your pictures on your computer, you can use the Library module to browse and organize them. To select the Library module, click **Library** at the top of the screen, or press **G**. These links (Library, Develop, Map, Slideshow, and Print) select the different Lightroom modules.

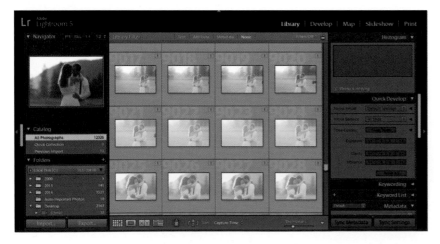

The sections that follow will give you an overview of the Library module.

THE GRID

The grid shows thumbnails of your pictures, as shown in the following figure.

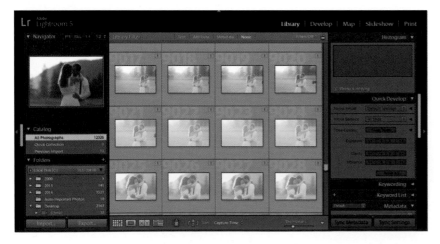

If you don't see your pictures in a grid, click Library at the top of the screen (as shown in the following figure) or press the G key. You can also click the Grid icon at the bottom of the Library module.

In the same area as the Grid icon, you can change the sort order and adjust the size of the thumbnails. Use small thumbnails to make it faster.

SELECTING PICTURES

Selecting a single picture is easy; just click it with your mouse. At times, you'll need to select multiple pictures to compare them side by side or apply settings to all of the pictures at once.

To select individual photos, hold down the **Ctrl** key on a PC or the **Cmd** key on a Mac and click each photo. Lightroom will select each individual photo.

To select a range of photos, click the first picture and then hold down the **Shift** key and click the last picture. Lightroom will select the first and last picture, and every picture in between.

To select all the pictures currently visible, press **Ctrl+A** on a PC or **Cmd+A** on a Mac. To clear your current selection, press **Ctrl+D** on a PC or **Cmd+D** on a Mac.

LOUPE

A loupe is an eyeglass that film photographers use to inspect an individual negative or slide. In Lightroom, it's just a preview of an individual picture, as shown in the following figure. The easiest way to view the Loupe is to double-click a picture. To return to the grid view, double-click the picture again, or press **G**.

By default, Lightroom fits the picture onto the screen. If you want to zoom in to see individual pixels, just click the picture once more. Once you've zoomed in to 100%, you can use the cursor to drag the picture around to see different parts of it. To drag with the cursor, simply click and hold the mouse button, and then move your cursor.

COMPARE

Frequently in photography, you must examine two pictures to determine which is better. To compare two pictures, click the first picture, and then **Ctrl-click** or **Cmd-click** the second picture. With two pictures selected, press **C** or click the Compare icon at the bottom of the Library module. Lightroom shows the pictures side by side.

Now you can click either picture, and they'll both zoom in, just like the Loupe view. If you want to zoom or pan them separately, unlock the lock icon at the bottom of the screen, as shown next.

When you've decided which of the two pictures you like better, click the white flag icon. If you don't like a picture, you can flag it for deletion by clicking the flag with an **X** on it.

You can also rate pictures from 1 to 5 stars by clicking the dots below the picture, as shown in the following figure. Later, you'll learn how to find the pictures that you've flagged or rated.

SURVEY VIEW

In addition to the Grid, Loupe, and Compare views, the Library module offers the slightly-less-useful Survey view. The Survey view shows you any number of pictures as big as possible. It's a bit like an on-screen contact sheet.

The Survey view is great for helping you choose between more than two pictures. If you just need to choose between two pictures, the Compare view is easier. To use the Survey view, first select the pictures that you want to view by **Ctrl-** or **Cmd-clicking** them. Then, press **N** on your keyboard or click the Survey View icon (as shown in the previous figure).

THE QUICK DEVELOP PANEL

You can fix many color and exposure problems using the Quick Develop panel, as shown next. Often, you can simply click **White Balance** and then click **Auto** to fix common color problems, such as pictures that are too orange or too blue.

If a picture is too dark or too bright, use the Exposure buttons, as shown next. Use **Clarity** and **Vibrance** to add some "pop" to your pictures. If you make your picture worse, just click **Reset All**. Every change in Lightroom can be undone; Lightroom makes no permanent changes to your pictures, so you never have to worry about it.

UNDOING YOUR MISTAKES

You can also undo mistakes one action at a time. Using a keyboard, press **Ctrl+Z** on a PC or **Cmd+Z** on a Mac. Or, with a mouse, open the **Edit** menu and select **Undo**. You can repeat the Undo command many times to go way back in time.

If you accidentally undo something, you can redo it by pressing **Ctrl+Y** on a PC or **Cmd+Y** on a Mac. Or, select **Edit | Redo** from the menu.

FIXING MANY PICTURES AT ONCE

At some point, every photographer takes a series of important pictures with the wrong camera setting. Perhaps you adjust exposure compensation on your camera to make your pictures darker and then spend an hour photographing your kid's soccer game. Or maybe you set your camera's white balance to cloudy and took 100 pictures at a friend's birthday party under florescent lights.

You can't go back in time and fix your camera settings, but you can use Lightroom to improve an entire series of pictures with just a few clicks.

From the Grid view, select multiple pictures using **Ctrl/Cmd** or **Shift-clicking**. Then, use the buttons on the Quick Develop panel to improve your shot. The changes you make will automatically apply to all the pictures you've selected, but it might take a few seconds.

If you've made changes to one picture and you want to apply them to many pictures, select the first picture and press **Ctrl+Shift+C** on a PC or **Cmd+Shift+C** on a Mac. This opens the Copy dialog box, shown next. You can access the Copy feature from the menus at **Photo | Develop Settings | Copy Settings**, but learning the keyboard shortcuts will definitely save you time.

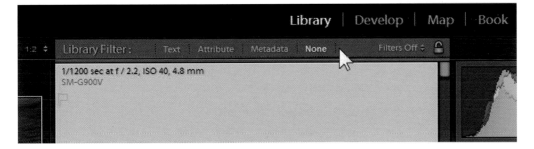

From the Copy dialog, select the types of edits that you want to transfer to other pictures. You might not be familiar with all the settings right now, but that's okay—just click **Check All** and then click **OK**.

Now, Lightroom has copied your edits to the clipboard. Select the picture or pictures, and then press **Ctrl+Shift+V** on a PC or **Cmd+Shift+V** on a Mac to apply those edits to the newly selected pictures. In a moment, Lightroom applies all the changes.

FINDING PICTURES

At some point, you'll have so many pictures that it's difficult to find that one special picture you're looking for. Lightroom's Filter Bar makes it easier to find them. The Filter Bar appears at the top of the Grid view, as shown next, so press **G** on your keyboard if you don't see the Grid.

If you see the Grid but don't see the Filter Bar, press the \ key, or select **View | Show Filter Bar** from the menu.

The Filter Bar shows three different types of filters by default: Text, Attribute, and Metadata. The next sections discuss each of these types of filters.

TEXT

Select the Text filter to search your pictures for a filename or keyword. This book discusses adding keywords to pictures later, but if you keyword all your children's pictures with their names, typing their name into the Text filter will show you every picture you've taken of that particular child. It's amazingly helpful, but only if you've already organized your pictures.

ATTRIBUTE

The Attribute filter shows pictures that match flags, star ratings, or colors that you've assigned within Lightroom. You can also filter pictures to only show still images or video.

As with the Text filter, the Attribute filter is only useful once you've already organized your pictures. If you use flags and stars to separate your best pictures, you can search for five-star pictures and quickly see your portfolio.

METADATA

If you haven't yet organized your pictures, the Metadata filter is the most useful. You can use the Metadata filter to find pictures taken on a particular day, month, or year, and narrow your results by the camera or lens you were using. If you happen to remember your camera settings for a specific picture, you can even use that as a filter. For example, if you remember you took a picture at 300mm and f/5.6, you can use the Metadata filter to show just those pictures.

You can change any of the columns to filter by different metadata. Simply click the column heading (such as **Date**, **Camera**, **Lens**, or **Label**) and then select the new metadata type, as shown next.

There are many other ways to find pictures, including using a map. We'll cover other methods throughout this book.

KEYWORDING PICTURES

If you're already excited by the idea of typing a name and finding every picture with that person or pet, you'll love keywording. Simply select all the pictures you want to add a keyword to, and then type a word in the Keywording panel, as shown next. To add more than one keyword, separate them with a comma.

Pictures can have as many keywords as you'd like to add, and Lightroom automatically adds frequently used keywords to the Keyword Suggestions list that you can click without having to touch the keyboard.

Once you have keyworded, use the Text filter to quickly find your pictures.

DEVELOPING PICTURES

The Develop module gives you more powerful editing tools than the Quick Develop panel. To select the Develop module, click **Develop**, or press **R** to jump to the Develop module and start cropping your picture (the first thing I usually do in the Develop module).

CROPPING AND ROTATING PICTURES

As I just mentioned, you can press **R** to crop a picture, no matter which module you've selected. You can also click the Crop icon to open the Crop & Straighten tool, as shown next.

Now, drag the corners of your picture to crop it.

By default, the crop tool maintains your picture's aspect ratio. The aspect ratio is the shape of your picture—how tall it is compared to how wide it is. If you'd like to make your picture square, or any other shape, click the Lock icon to unlock it, and then drag the corners freely.

The aspect ratio is also useful when printing pictures. For example, if you're going to make an 8x10" print, most cameras require you to crop the edges of the picture. If you send your full picture to a printing service, they'll randomly crop off the edges, and it might ruin your composition. To crop it yourself, click the **Aspect** list, and then select **8x10**, as shown next.

You can also use the Crop tool to rotate your pictures. Move your cursor outside the edge of the picture, and it will turn into the Rotate tool. Now, click and drag your cursor to straighten the picture. This is a great way to fix a horizon that isn't level.

REMOVING SPOTS

Dust on your sensor leaves spots in your pictures. Fortunately, Lightroom makes it easy to remove dust spots. First, select the Heal tool, as shown here.

Next, use the scroll wheel on your mouse to adjust the brush size to be slightly larger than the spot, as shown in the following figure.

Now, click the spot to remove it. If you need to remove something longer than a spot (such as a hair on your sensor), you can drag your cursor over the mark to paint over it.

Lightroom looks around your picture and tries to find a similar part of the picture to copy over the spot. Sometimes, Lightroom copies from the wrong part of the picture. If that happens, just grab the source circle and drag it to a part of the picture that better matches the spot you want to clean.

Sensor dust will appear in every one of your pictures. To quickly remove it from many pictures at once, copy and paste the settings from the first picture that you fix, as described in "Fixing Many Pictures at Once" earlier in this chapter.

EXPORTING AND SHARING PICTURES

You must export pictures from Lightroom before you can see any edits that you've made. You cannot simply use your pictures directly from Explorer or Finder; they won't have your edits.

To export pictures, first select them in the Grid view. Then select **File | Export** from the menu, or press **Ctrl+Shift+E** on a PC or **Cmd+Shift+E** on a Mac. This opens the Export dialog, as shown next.

The Export dialog has far too many options. However, all you really need to do is to click the **Choose** button and select a folder to save your exported pictures into. The default settings are fine for most.

Then, click **Export** to save your picture as a JPG that you can upload to Twitter, Facebook, or a printing service. Once you have exported the pictures, you can select the folder you chose in Explorer or Finder to use them.

SAVING DISK SPACE

Lightroom generates preview files that make it faster to browse your pictures. Over months and years, these previews can consume a great deal of disk space.

If you run low on disk space, you should remove the previews to save space. From the menu select **Library | Previews | Discard 1:1 Previews**. You might also want to select **Discard Smart Previews** from the same menu. We'll cover previews later in this book.

SUMMARY

Most casual photographers won't need to use all of Lightroom's features. In this chapter, I've tried to show you the most important features so you can get started organizing and editing your own photos.

I've only scratched the surface of what Lightroom can do. When you're ready, read on, and you'll learn how to save time and make even more stunning photos.

2
Basic Adjustments

To view the videos that accompany this chapter, scan the QR code or follow the link:

SDP.io/LR5Ch2

The Library module provides basic editing tools, allowing you to instantly fix many exposure, color, and contrast problems. While the editing tools in the Develop module are more powerful, learning the Library module adjustment tools can save you a few clicks per picture. That can add up over the thousands of pictures that you'll take in a lifetime.

Even if you prefer to use the Develop module, this chapter provides an important overview of the histogram and how to use white balance, color temperature, tint, exposure, contrast, highlights, shadows, whites, blacks, clarity, and vibrance.

USING THE HISTOGRAM PANEL

At the top of the right pane, you'll see the Histogram panel, which shows you how much brightness and darkness the selected picture has. You might think, "That's pointless; I can see the picture with my eyes." If that's the case, you can simply hide the Histogram panel by clicking the triangle, as shown in the next figure.

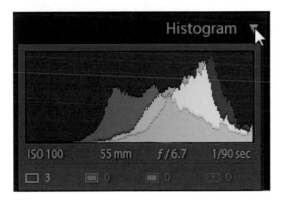

Hiding the Histogram panel is a terrible idea, however, and here's why: your monitor can lie to you, but a histogram always tells the truth. If you're in a brightly lit room, or if there's a bright window behind your monitor, your pictures might look darker than they really are. Or, if you're in a dark room, and you have your monitor's brightness turned up, your picture might look brighter than it is. When you print or share that picture, people will see it in a different environment, and they'll see how bright or dark the picture truly is.

In other words, you can't trust your eyes because your monitor, your brain, and the light in the room can make your picture look brighter or darker than it really is. Instead, you should trust the histogram.

A histogram is a bar chart showing you how much of your picture is dark or bright. Every completely black pixel in your picture is stacked up on the left side of the histogram. If you have a totally black picture, the histogram will look like a mountain pushed against the left.

Every completely white pixel in your picture is stacked up on the right side of the histogram. If you overexposed a picture to make it mostly white, it would look like a mountain pushed up against the right.

In between total black and total white, we have shadows (near the left side), mid-tones (in the middle), and highlights (on the right).

Here's the histogram of a normally exposed picture. As you can see, it peaks in the middle (the mid-tones). Even without looking at the picture, the histogram tells you that it's well-exposed, with both shadows and highlights. The peaks at the left of the histogram are the darkest parts of the picture, such as the sunglasses and necklace. The peaks at the right side of the histogram represent the bright background. The blue spike in the histogram represents the blue dress, and the yellow spike represents the yellow tent in the background.

Here's the histogram of a dark (low-key) picture. Though the picture is properly exposed, the trees are in shadow, and many of the outfits are dark. Because so much of the picture is dark, the histogram pushes against the left side of the frame.

Here's the histogram of a bright (high-key) picture. As you can see, the histogram is mostly pushed towards the right. It's also a very warm picture, meaning it has more orange and red tones than blue and green tones. The histogram shows that, too, because the orange and red parts of the histogram are higher. If you didn't intend the picture to be bright, this histogram would indicate an overexposed picture.

For this next example, I used the Quick Develop panel to drop the exposure one stop. As you can see, Lightroom made the picture darker, and the histogram shifted to the left.

Practice: Go through some of your pictures and see what the histogram looks like. Try adjusting the exposure up or down, and see how the histogram changes.

Almost every picture should have some white and some black in it. Even if you take a night photo, the histogram should stretch all the way to the right side. If a histogram doesn't stretch to the left and right sides, it's low contrast, and might look bad. Therefore, you should adjust the exposure in the Quick Develop panel or use the Develop module (discussed later in this book) to adjust the contrast.

That's what I mean when I say to trust the histogram; a picture might look fine on your monitor, but if you see that the histogram doesn't touch the right side, it's probably underexposed or low contrast. If it doesn't touch the left side, you should probably adjust it so that some part of the picture is pure black.

These are just general guidelines; I would never want to stifle your creativity. Make all the high-key or low-key photos you want, but do so deliberately. Too often, I see photos that are unknowingly dark, bright, or washed-out simply because the photographer trusted their monitor instead of their histogram.

Using the Quick Develop and Basic Panels

The Quick Develop panel is appropriately named. It doesn't do much at all, but it can save you a click to the Develop module. The Develop module provides the same functionality using the Basic panel. Let's take a look at the settings one-by-one, from top to bottom.

Saved Preset

Presets are a group of settings that can change the brightness, color, sharpness, or many other aspects of your picture. Lightroom includes several presets; you can also create your own, and you can download more free from the Internet. You'll learn more about presets when you study the Develop module.

The next figure shows the author using a photographic technique known as forced perspective to embarrass his wife in Paris. By using the Red Filter saved preset, I was able to instantly convert it to black-and-white.

Practice: Try using the Quick Develop panel to apply different presets to your pictures. If you don't like a preset, you can press **Ctrl+Z** (on a PC), or **Cmd+Z** (on a Mac), or click **Reset All** to undo your changes.

While you can use the Quick Develop panel to apply presets, it's usually easier to use the Develop module. In the Develop module, you can see a preview of the preset before you apply it, and browsing the presets is easier.

Adobe has hidden two additional settings underneath **Saved Preset**. You can access the **Crop Ratio** and **Treatment** settings by clicking the Triangle beside Saved Preset. The Crop Ratio changes how tall or wide a picture is, allowing you to switch it to square or panoramic. For more information about cropping, refer to Chapter 5. The Treatment setting allows you to switch a picture from color to black-and-white. For detailed information about using black-and-white, refer to Chapter 9.

There's no good reason to ever use these hidden settings; you should use the Develop module instead. If you were to apply a different crop ratio, you would have to switch to the Develop module to move the crop, anyway. If you were to convert a picture to black-and-white, you would have to switch to the Develop module to control the brightness of the individual colors.

WHITE BALANCE

White balance controls a picture's colors, and allows you to fix pictures that have an ugly tint to them. For example, if an indoor picture looks too orange, you can use the **White Balance** tool to fix it.

This next example shows a picture I took inside a church in Istanbul. By simply looking at the histogram, you can see that it's a very warm picture, with more reds and yellows than blues and greens. This was caused by the incandescent light in the church, which is very orange. Artificial light appears orange to a camera, even though your brain might see it as white. As a result, the picture has an artificial tint.

I can let Lightroom try to automatically fix the white balance problem by selecting **Auto** White Balance, as shown next. Lightroom did decrease the yellows and increase the blues, providing more natural white balance. You can see this reflected in the histogram, too; the blues moved to the right, while the yellows moved to the left.

You can also manually select different presets, including **Daylight**, **Cloudy**, and **Florescent**. However, you might find it faster simply to switch to the Develop module, as discussed later in this book.

Practice: Find a picture with bad color and change the **White Balance** from **As Shot** to **Auto**. Does it improve it?

If you click the White Balance triangle, you can use the **Temperature** and **Tint** buttons to manually adjust the color. Temperature shifts the white balance from blue (on the left) to orange (on the right). Tint shifts the white balance from yellow (on the left) to purple (on the right). The next figure shows the same photo with the Tint adjusted all the way to the right. The Develop module has better tools for manually adjusting the white balance, so you shouldn't ever need to use these buttons.

Switching over to the Basic panel in the Develop module, the white balance dropper can quickly solve most color problems. Click the dropper (shown next), and then click part of the picture that should be white or grey.

Lightroom shows a close-up view of the pixels that the dropper will sample. Once you click part of the picture, Lightroom will adjust the entire picture so that the pixels you clicked are perfectly white.

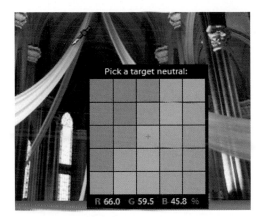

Many perfectionist photographers carry a grey card or a white card with them, and they take a test photo of the card in the environment's light. Then, they'll use the dropper to set the white balance off the card, knowing that the card is perfectly white. This gives them perfect color every time. This extra step never hurts, but it's rarely necessary in modern photography for three reasons:

- There's almost always something white or grey in a photo to set the white balance from.
- Tools like Lightroom simplify fixing white balance settings in post-processing.
- Technically accurate colors are often not a priority; most published pictures have some toning effects applied. In other words, colors often don't have to be accurate to be good.

TONE CONTROL

A picture's tone is its brightness and contrast, and the Quick Develop module allows you to easily make adjustments. Whereas Adobe punished you for clicking the Saved Preset and White Balance triangles by showing you buttons that you should instead ignore, the Tone Control triangle reveals a set of very useful buttons, as shown in the following figure.

You can also adjust these settings using the Basic panel in the Develop module, shown next. They both do exactly the same thing, but the Develop module has sliders rather than buttons.

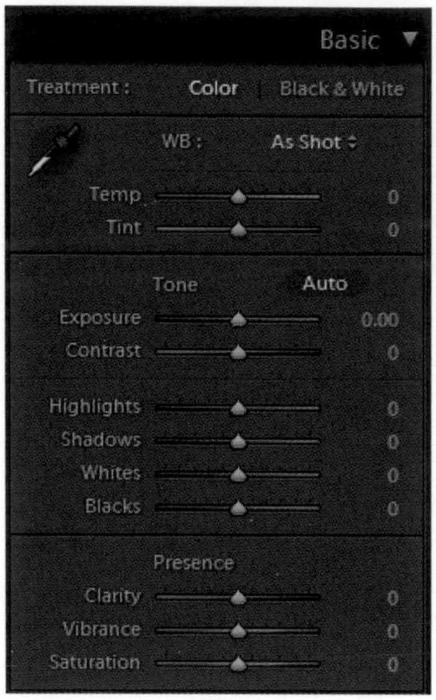

The easiest way to understand what each of them does is to simply experiment, but I'll give you a brief overview in the sections that follow.

WAIT—THAT'S NOT WHAT I SEE!

If you see different controls under Tone Control, it's because the picture you've selected is using a different generation of Lightroom's processing. For example, if your Tone Control panel looks like the following example, your picture is set to use Lightroom's outdated 2003 process. Notice that it has **Fill Light** instead of **Shadows**, and **Recovery** instead of **Highlights**.

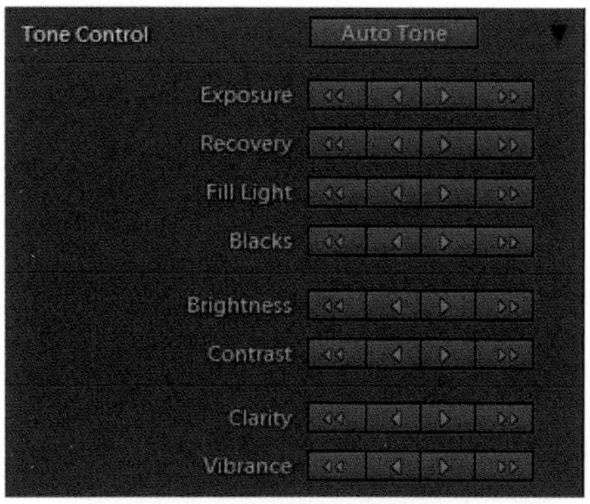

Lightroom does this for backwards-compatibility with earlier versions. To fix this, switch to the Develop module, expand the Camera Calibration panel (at the bottom-right), and change the Process to **Current**.

AUTO TONE

The **Auto Tone** button adjusts the exposure and contrast so that a picture has both bright whites and dark blacks. It's useful to quickly improve many pictures, but it'll ruin pictures that are deliberately high or low key. The following example shows how **Auto Tone** adjusted a slightly underexposed snow photo.

EXPOSURE

Click the right **Exposure** buttons to make a picture brighter, or click the left buttons to make a picture darker. It's better to get the exposure right in-camera, but these buttons can help.

CONTRAST

Contrast is the separation between light and dark. If a picture is washed-out, increasing the contrast can fix it. You might want to decrease the contrast of a picture taken in direct sunlight.

The following example shows the effect of decreasing the contrast in a very contrasty picture. As you can see, the dark foreground is much brighter than the original on the left, and the bright background is less bright.

These two histograms also demonstrate the difference, with the original again on the left. As you can see, the peaks on the left and right sides are pulled into the middle when you reduce contrast.

HIGHLIGHTS

Highlights are the brightest parts of your picture. The left-pointing **Highlights** buttons make the brightest parts of your picture darker, reducing contrast. The right-pointing **Highlights** buttons increase contrast by making the bright parts even brighter (without making them completely white).

If the sky is too bright in a picture, you might try using the left **Highlights** buttons to fix it. For example, clicking the **Highlights** << button four times reduced the brightness of the background compared to the original picture on the left. It left the shadowy foreground untouched, however.

The next examples show the histograms for these pictures, starting with the original. As you can see, adjusting the highlights impacted (and almost eliminated) the right part of the histogram. The shadows on the left of the histogram are virtually unchanged.

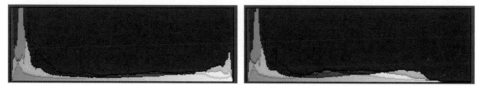

SHADOWS

Shadows are the darkest parts of your picture. Clicking the left buttons will make them even darker, without making them completely disappear into blackness. Clicking the right buttons will show more shadow detail by turning the shadows into mid-tones.

The following example shows the effect of clicking the **Shadows** >> button three times. As you can see, the bright background is completely unchanged compared to the original on the left, but the dark foreground is much brighter.

By comparison, clicking the **Shadows** << button three times hides most of the shadow detail, creating a silhouette.

WHITES

Clicking the right **Whites** buttons increases both the overall brightness of your picture and the overall contrast. It can be a bit confusing to understand the difference between the **Whites** adjustments and the adjustments for **Exposure**, **Contrast**, and **Highlights**. Indeed, they can create similar effects, but with important differences.

The right **Whites** buttons makes your entire picture brighter by making some highlights completely white. If you click the right **Whites** buttons while watching the histogram, you'll see that some of the highlights become completely white, while the entire histogram slides to the right just a bit.

You should use the right **Whites** buttons if a picture is generally well-exposed, but no part of the picture is completely white. The following before/after example shows how pushing the **Whites >>** button once made a picture overall brighter and more contrasty, but particularly brightened the highlights in the bird.

To better illustrate the effect of **Whites >>**, press it multiple times, as the following example shows. All highlights in the picture became completely white, hiding the detail in the egret.

The following figures show the histograms of the image before and after applying five clicks of the **Whites >>** button. As you can see, the histogram seems to have moved to the right, indicating that the entire image is brighter. As parts of the image approached the right edge of the histogram, more of the picture was filled with bright white, making the histogram climb up the right side.

You should only use the left **Whites** buttons if you took a picture using raw, rather than JPG. If you use the left **Whites** buttons with a JPG image, the brightest parts of your picture will simply become grey, reducing contrast. If you use the left Whites buttons with a raw image, you might be able to recover some overexposed highlights. For more information about using raw images, refer to Chapter 4 of *Stunning Digital Photography*.

BLACKS

The **Whites** button clips or recovers the brightest parts of the picture and distributes the rest of the image across the entire histogram. The **Blacks** buttons clip or recover the darkest parts of the picture. For example, if you click the left **Blacks** buttons, the entire picture becomes darker, and some of the shadows will become completely black.

The following example shows three clicks of the **Blacks** << button. The shadows of the original picture (shown on the left) are completely black in the after picture (shown on the right). The mid-tones of the blue sky are noticeably darker. The highlights are also a bit darker, but the brightest white parts of the picture are unchanged.

You should only use the right **Blacks** buttons if you took a picture using raw, rather than JPG. If your image is raw, Lightroom will attempt to recover the shadows and brighten the entire picture.

CLARITY

Clarity adds punch to your pictures. It adds some contrast to mid-tones and brings out edges. The easiest way to understand it is to experiment with it. Be gentle when increasing clarity; the single most common post-processing problem I see is overdone clarity.

This example shows a cardinal before-and-after with three clicks of **Clarity** >>. As you can see, the picture seems to have much more contrast. In particular, notice the individual feathers on the cardinal stand out much stronger.

The following example shows the effect of three **Clarity** >> clicks on a portrait. As you can see, the shadows become much deeper, creating the effect of greater contrast and faster light fall-off.

VIBRANCE AND SATURATION

Vibrance intelligently increases the saturation of colors. If a picture seems dull or washed out, clicking the right buttons might improve the color. Be careful with this, though, because it's tempting to overdo it. If you click the right buttons three times and are happy with the result, click the left button once to back off the vibrance a little; the rest of the world will probably be happier with a more natural image.

You might ask, "What's the difference between the Vibrance and Saturation sliders in the Develop module?" Here's a clue: the Vibrance setting was important enough to make it to the Quick Develop panel, but Adobe chose to leave Saturation out. Saturation increases the intensity of all the colors in your image. Vibrance is smarter, because it primarily increases the dullest colors in your picture, while being very gentle with the colors that are already bright.

The following example shows the cardinal picture with three clicks of **Vibrance** >>. Notice that the yellows and reds are a bit more saturated, but because they're already very saturated, Vibrance didn't adjust them much. The blue sky, however, was not very saturated in the original picture. Therefore, the Vibrance adjustment made a much bigger difference.

The next example shows the same picture unedited, with +60 Saturation (applied using the Develop module) and with +60 Vibrance (equal to three **Vibrance** >> clicks). As you can see, the image with increased saturation loses details in the cardinal, because he's already very saturated. The image with +60 Vibrance still shows all the details of his feathers and overall looks much more natural, even though the saturation of the blue sky is far greater.

USING VIRTUAL COPIES

Sometimes, you might want to process a single image multiple different ways. For example, you might want to crop an image for horizontal use on the web, or vertical use in print. Or, you might want to create a black-and-white version of an image while still having access to the full-color version.

Virtual copies make this easy. Virtual copies don't literally create a second copy of your image; that would waste disk space. Instead, they create a second copy of the record of the image in Lightroom, allowing you to process it entirely differently.

Right-click an image and click **Create Virtual Copy**. Lightroom shows a second copy of the image. You can repeat the process to create as many virtual copies as you want. I often make a dozen virtual copies of an image when I experiment with different processing looks.

The virtual copy includes all of the edits you've already made to the original image, but if you make future edits to the original image, they won't be applied to the virtual copy. You can recognize a virtual copy in the grid view or filmstrip by the folded corner.

USING STACKS

When you create a virtual copy, Lightroom adds the new copy to a stack with the original photo. This doesn't change how you use either image; they're completely separate. You can recognize a stack by the numbers written in the upper-left corner, as shown next. To make it easier to browse photos, you can expand or collapse a stack by double-clicking the numbers.

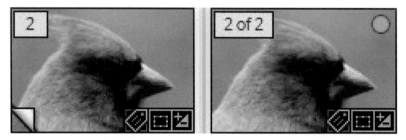

Lightroom stacks virtual copies by default. If you edit an image with Photoshop or another external editor, Lightroom might make a second copy of your original image. In that case, it will also stack the second copy.

You can stack any two pictures, which is useful when you want to reduce the clutter in your grid view. It's particularly useful to stack bracketed photos or photos that are part of a time lapse, since the photos all belong together. To stack images, select them, right-click one image, select **Stacking**, and then select **Group Into Stack**. That shortcut menu also allows you to unstack images and collapse or expand the stack.

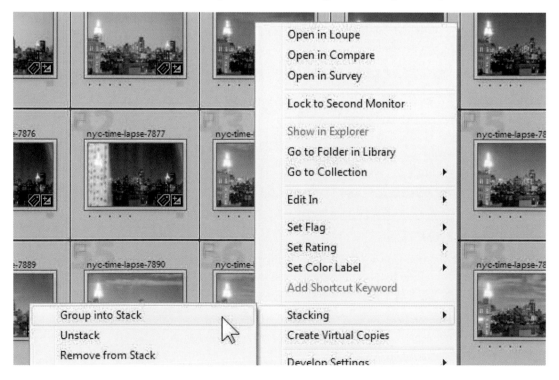

Lightroom has a fairly unintelligent tool that automatically stacks images based on capture time. If two pictures were taken within a certain number of seconds of each other, Lightroom will put them into a stack.

To open the tool, select the images you want to consider for stacking and select **Photo | Stacking | Auto-Stack By Capture Time**. Adjust the maximum time between images, as shown next, and then click **Stack**.

The Auto-Stack tool has potential, but in practice, it's not as useful as you might hope. It works as expected, but in the real world, photographers do take completely unrelated pictures within a few seconds of each other. When that happens, Lightroom will stack those unrelated images, and if the stack is collapsed, you might completely overlook an important image.

To me, the risk of hiding an important picture outweighs any time-saving benefits of the Auto-Stack tool. However, it is useful in specific scenarios.

SUMMARY

Lightroom's Histogram and Quick Develop panels are useful tools for fast edits and don't require you to jump to the Develop module. With these tools, you can fix underexposed and overexposed pictures and even washed-out pictures. For more serious editing, including cropping, you'll need to switch to the Develop module (discussed later in this book).

3
Organizing Your Pictures

To view the videos that accompany this chapter, scan the QR code or follow the link:

SDP.io/LR5Ch3

The Library module is how you'll organize and find your pictures. It's extremely powerful, catering to almost any structure you can imagine, whether you are using keywords, colors, folders, or ratings. It's good enough for professionals who need to re-print wedding photos from a decade ago, and it's also perfect when you want to find your child's first-day-of-school pictures from every grade.

This chapter digs deep into the Library module, and it assumes you've read Chapters 1 and 2. For many, it will be more detail than you need. Feel free to skip sections that don't seem useful to you, or to skip the entire chapter and watch the video instead.

In the sections that follow, we'll describe each of the organizational tools in Lightroom.

KEYWORD SETS

As described in Chapter 1, you can add keywords to pictures by selecting them and then typing keywords in the Keywording panel. If you're frequently typing the same keywords, Lightroom's Suggested Keywords will work perfectly. However, if you regularly use more than a dozen keywords, and you switch between different groups of keywords for different types of photos, keyword groups can help.

A Keyword Set is a group of nine related keywords. From within the Keywording panel, you can select one of the built-in keyword sets or create your own. For example, a husband might create a keyword set named "my family" with the names of his relatives, and a set named "in-laws" with the names of everyone in his wife's family. That could speed up the tagging process.

Frankly, most people don't need to use keyword sets, but it's nice that Lightroom provides features for different types of photographers. It's also nice that Lightroom makes it easy to hide features you don't need.

USING THE KEYWORD LIST PANEL

The Keyword List panel is just a list of every keyword you've used. It's much better to select keywords from the Keyword List than to re-type them, because it eliminates the possibility that you'll make a typo. It also prevents you from tagging some pictures of the same person as 'Kate,' 'Katie,' and 'Katherine.'

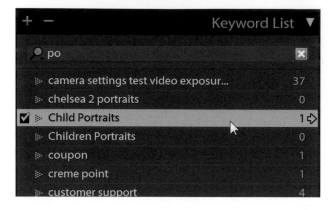

The Keyword List is about consistency, and consistency is critical for accurate data entry. Besides improving accuracy, the Keyword List panel also allows you to create synonyms for keywords and to prevent specific keywords from being exported.

Creating synonyms for keywords allows you to add multiple keywords with a single click. For example, Glacier National Park is in the state of Montana in the United States. Therefore, I might want to add synonyms for the Glacier National Park keyword as shown in the next example. Now, every picture that's tagged with 'glacier national park' will appear in my search results when I filter for 'Montana,' 'MT,' 'United States,' or any other synonym—even if I haven't updated the tagged pictures.

Nicknames are a great place to use keyword synonyms. To make sure you can find pictures of your friend Katie, whom you call Kate when you're formal and Katherine when you're angry, you could add Kate and Katherine as synonyms of Katie. Then, pictures tagged with Katie will appear when you search any of the synonyms.

Notice the three Keyword Tag Options in the previous dialog:

- **Include On Export.** Clear this checkbox to prevent this keyword from being written to the metadata of pictures that you export. Note that this can be overridden if a synonym exists and has **Export Containing Keywords** selected.

- **Export Containing Keywords.** Select this checkbox and Lightroom will add the parent keyword (in the previous example, 'Glacier National Park') to the metadata when exporting a picture that has a synonym as a keyword (such as 'Montana').

- **Export Synonyms.** Select this checkbox and Lightroom will automatically add all synonyms to the metadata when exporting a picture that has the selected keyword. In the previous example, if a picture is tagged with 'Glacier National Park,' Lightroom will add 'Montana,' 'MT,' and all the other synonyms to exported pictures.

Generally, the more keywords, the easier it is to find your pictures. Therefore, I usually recommend leaving all the checkboxes selected. If you have keywords that you don't want to be public, such as, 'babies who pooped during the shoot,' or 'asked me to Photoshop them thinner,' then you should clear the checkboxes. But don't use keywords that you don't want to be public, because it's too easy to make a mistake.

USING THE METADATA PANEL

The Metadata panel reveals all that gorgeous data that your camera stores in the pictures, including your aperture, shutter speed, and ISO. It also allows you to add your own data, including the Title, Caption, and Copyright.

Some of the fields, such as **Title** and **Caption**, can be edited. These fields can be searched, but keywords are usually better for words that you want to use to find pictures. Instead, you should use Title and Caption for their intended purposes, and they'll show up in interesting places.

For example, if you create a slideshow (using the Slideshow module), Lightroom can, optionally, display the Title and Caption. If you create a photo book (using the Photo module), Lightroom can show that information in the book.

The Title and Caption will be exported by default and can be used by external applications. For example, if I export a picture and upload it to 500px (a photo sharing website), 500px will automatically name my picture with the Title I assigned, so I won't have to re-type it. Setting the Title and Caption in Lightroom can save a great deal of time, especially if you upload your pictures to multiple websites.

If you don't set the **Copyright** field, don't worry. In the United States and most countries, you automatically own the copyright to any picture you take, even if you don't type your name. However, if someone finds a picture online and wants to track down the photographer, they might use the field to find your name (but few will go through the trouble).

Notice that the Metadata panel is pretty tall by default. If it takes up too much space, you can switch from the default set of metadata to something more specific. For example, clicking the **Metadata** list and then selecting EXIF shows only your camera settings, hiding information that you might not use, including Title and Caption.

RATINGS

Ratings are the most useful feature in Lightroom, allowing you to assign pictures 1-5 stars. 1 star, of course, represents a terrible picture, while 5 stars represents your best work. If you rate your pictures, you can use a Text filter to quickly find the 43,000 pictures you've taken of your cat and then use an Attribute filter to narrow it down to just your 5-star pictures.

In the Grid view, you can click the star rating you'd like to apply to a picture, as shown in the next example. However, it's faster to use the keyboard; just press a number from 1-5 to assign all currently selected pictures a star rating.

Here's the system I use for rating my pictures:

★   This is a picture with technical problems, such as a blurry picture. If I run out of disk space, I can delete 1-star pictures without feeling bad about it.

★ ★   This is a picture that is either bad or a duplicate of another picture. I don't ever plan to delete it, however. I use 2 stars to indicate that I shouldn't ever bother to look at a picture, though I don't have the heart to permanently delete it.

★ ★ ★   3-star pictures are decent, but not striking.

★ ★ ★ ★   4-star pictures are great pictures, but they're not quite portfolio worthy. When I'm sorting my pictures, 4 stars is the highest I generally choose.

★ ★ ★ ★ ★   5-star pictures are my very best pictures, the ones I consider worth of putting into my portfolio. I try not to select more than a few dozen 5-star pictures a year.

COLOR LABELS

Another attribute type useful for organizing your pictures is colors labels. You can assign one of six colors to your photos, and then see at a glance which color a photo is. For example, you might assign all your personal photos the color red, and all your professional work the color yellow. Or, you might assign wildlife photos green and family photos blue.

You can assign color labels using the Set Color Label right-click menu, but that would be a slow process. Keyboard shortcuts are a faster way to assign color labels:

- ■ 6: Red
- ■ 7: Yellow
- ■ 8: Green
- ■ 9: Blue

Notice that there are six color labels, but only four keyboard shortcuts. 0-5 are used to assign star ratings, which are generally more useful than color labels because they can be sorted.

If you do use color labels, I suggest making your own label presets. Rather than naming the colors Red, Yellow, and Green, you can change them to your own custom meanings, such as Family, Business, and Pets. From the menu, select **Metadata | Color Label Set | Edit**.

Use the Edit Color Label Set dialog to rename the colors. Once renamed, click the **Preset** list and select **Save Current Settings As New Preset**, as shown next.

Edit Color Label Set

Preset: Lightroom Default (edited)

- ✓ Lightroom Default (edited)
- Bridge Default
- Lightroom Default
- Review Status
- Save Current Settings as New Preset...
- Restore Default Presets
- Update Preset "Lightroom Default"

● Portrait

○ Landsca

● Wildlife

● Sports

● Macro

If you wish to maintain compatibility with labels in Adobe Bridge, use the same names in both applications.

Change   Cancel

It gets really confusing if you switch between multiple color label sets, so I suggest using a single color label set for all your catalogs. If you do create multiple sets, you can switch between them using the **Metadata | Color Label Set** menu.

FLAGS

As discussed in Chapter 1, flags are one of the most useful ways to organize your pictures. As you're flipping through a new set of pictures, press **P** on the pictures you want to edit further, or press **X** when you find a picture you should just delete.

With your pictures flagged, you can use filters to quickly find the pictures that need more editing. When you need to free up disk space, use the **Photos | Delete Rejected Photos** menu item.

USING THE FILTER BAR

Chapter 1 showed you how to use filters to quickly find your pictures. The Filmstrip pane at the bottom of the Lightroom window provides shortcuts to creating filters, allowing you to quickly find your pictures. It's self-explanatory; I just wanted to point it out.

PAINTING

You can use painting to apply metadata, settings, and attributes between different pictures (but you probably don't want to). First click the Painter icon at the bottom of the Grid view, as shown next.

Now, select the settings and value that you want to apply to your pictures. For example, if you select **Flag**, you will need to choose whether to apply **Flagged**, **Unflagged**, or **Rejected**. If you select **Rating**, you need to select whether to apply 1-5 stars.

Painting seems like an absurd way to apply settings to multiple pictures. I find it faster and easier to simply select multiple pictures and then make the adjustments as I normally would.

SUMMARY

Without exaggeration, Lightroom is the world's most powerful and flexible way to organize your photos. Few of us will choose to use all of Lightroom's organizational tools, but even if you just rate your pictures and add keywords, you'll have a much easier time finding your pictures.

4
Collections & Folders

To view the videos that accompany this chapter, scan the QR code or follow the link:

SDP.io/LR5Ch4

Collections are the best way to group your pictures into hierarchy. If you're tempted to make folders to organize your pictures, don't. Use collections instead. Collections are like virtual folders, but they're much smarter.

Of course, Adobe found a way to make collections complicated enough that I need to explain the different types to you:

- **Collection Set.** Collection sets don't store pictures; they store collections or other collection sets. Collection sets are strictly useful for organizing your collections. At some point, you'll get more than twenty collections, and you'll be tired of scrolling through them. Create a few Collection Sets to organize your collections, and drop your collections into them.

- **Collection.** Collections store pictures that you manually add to them. You can put collections into collection sets, but not into other collections.

- **Smart Collection.** Smart collections automatically find pictures in your library based on criteria you specify. For example, you could create a smart collection that matches all pictures taken with your 70-200 lens, or all your 5-star pictures. You can't manually add pictures to a smart collection.

Pictures can be in more than one collection at a time. If you make edits to a picture in a collection, they change the picture when you view it outside the collection, too. Therefore, you might want to make virtual copies when putting a picture into a collection.

CREATING COLLECTIONS AND COLLECTION SETS

The easiest way to create a collection is to select one or more pictures, right-click them, and then click the Collections plus symbol, as shown next. Collection sets can't store pictures, so it doesn't matter whether you have any pictures selected.

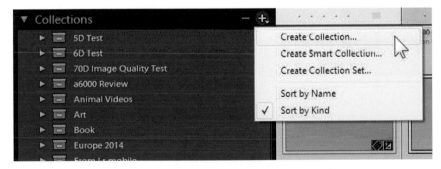

After selecting Collection, use the Create Collection dialog to name the collection and specify the options. If you select collection set, Lightroom just asks you the name and whether you want a containing collection set.

You have these options when creating a collection:

- **Inside a Collection Set.** Select this checkbox if you don't want this to be a top-level collection, and then select the Collection Set you want to store this inside.

- **Include selected photos.** If you selected pictures prior to creating the collection, select this checkbox. If you want to create an empty collection, clear this checkbox.

- **Make new virtual copies.** If you'd like to be able to make edits to the pictures in the collection without affecting your original pictures, select this checkbox. Lightroom will create a virtual copy of every picture you're adding to the collection.

- **Set as target collection.** You can specify one collection as the target collection and easily add other pictures to it. If you plan to add more pictures to the new collection, select this checkbox.

- **Sync with Lightroom mobile.** If you use Lightroom Mobile on your iPhone or iPad, select this checkbox to have your pictures synced. This requires a Creative Cloud subscription, and it doesn't work very well, so I recommend leaving this checkbox cleared.

CREATING SMART COLLECTIONS

Smart collections are a saved search, a quick way to look through your library for any pictures meeting specific criteria. Just as with collections and collection sets, you can create a smart collection by clicking the Collections Plus symbol, as shown next.

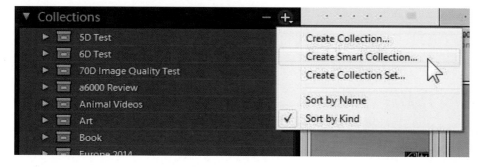

When you create a new smart collection or edit an existing smart collection, Lightroom shows you the Edit Smart Collection dialog box, shown next. Use this tool to name the smart collection and specify the search criteria.

MANY ALL, ANY, OR NONE

The most important option is the most easily overlooked: Match (All, Any, or None) Of The Following Rules. If you have more than one rule, Match All shows you only those pictures that meet every criteria you list, while Match Any shows you all pictures that meet at least one of the criteria you specify.

Select Match None to show you pictures that don't meet any of your criteria. For example, if you use a Nikon D3300 to take most of your pictures, you could create a smart collection that showed you every picture taken with any camera except your Nikon D3300. The next example demonstrates this.

You never need to use Match None, though, because you can simply choose the 'Is Not' operator when creating a criteria.

SPECIFYING CRITERIA

You can use just about any type of data to create a smart collection, including the camera settings, date, ratings, and camera gear. You can choose any of the following data types:

- **Rating.** The number of stars you've assigned a picture.

- **Pick Flag.** Whether you've flagged a picture. Create a Smart Collection for pictures flagged as rejected to quickly find and delete unwanted pictures when your disk space is low.

- **Label Color.** Specify the color you've assigned to a picture.

- **Label Text.** If you customized the names of your colors, you can search labels by their text.

- **Has Smart Preview.** If (for some reason) you want to find pictures that have a smart preview, use this criteria.

- **Source.** Find pictures contained in a specific collection or folder.

- **File Name/Type.** Find pictures with a phrase in the filename. You could use this to find all jpg files by searching for .jpg.

- **Date.** Find pictures taken on a specific date or date range. Use this to create Smart Collections such as, 'Best Pictures of 2014.'

- **Camera Info.** Select pictures based on the camera used or camera settings such as aperture and shutter speed.

- **Location.** If you've tagged your pictures, or your camera has GPS, use Location to find pictures taken in specific places.

- **Other Metadata.** Search pictures based on titles, captions, and keywords.

- **Develop.** Find pictures that have been edited or cropped.

- **Size.** One of the most useful data types is buried under size: Aspect Ratio. Create Smart Collections that find all your top-rated horizontal (landscape) or vertical (portrait) pictures, and you can quickly find a picture to fit that empty frame on your wall.

- **Color.** This sounds like it might let you find all your red or blue pictures, or at least identify pictures you've been editing to black and white. Sorry, the name is misleading, and this is perhaps the most useless data type. I have a hard time imagining photographers who frequently search for pictures based on their color mode, but this is that option if you need it.

- **Any Searchable Text.** The easiest way to find pictures based on filename, caption, or keyword. I recommend using this data type for creating smart collections for your daughter's, dog's, or dad's names.

To the right of the data type is the operator, which tells the smart collection how you want to compare the data. It's different for every data type.

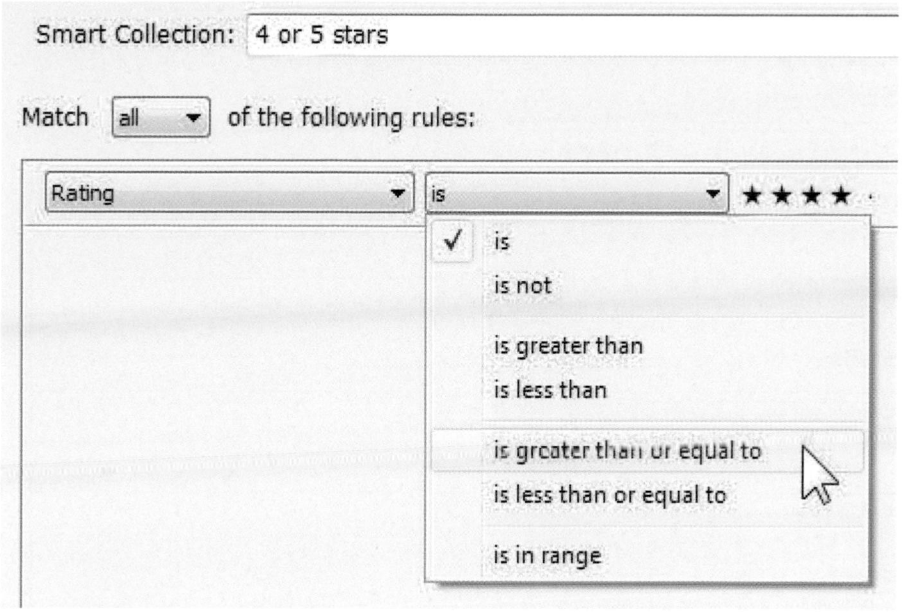

If you want to use multiple criteria to search for pictures, click the + button, as shown next. Click the – button to remove a criterion. When you have more than one criterion, remember to select Match Any or Match All carefully.

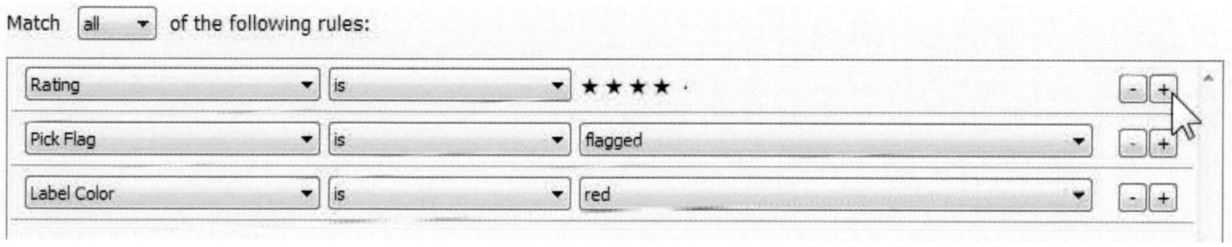

EXAMPLE SMART COLLECTIONS

Here are some example smart collections that you might want to use:

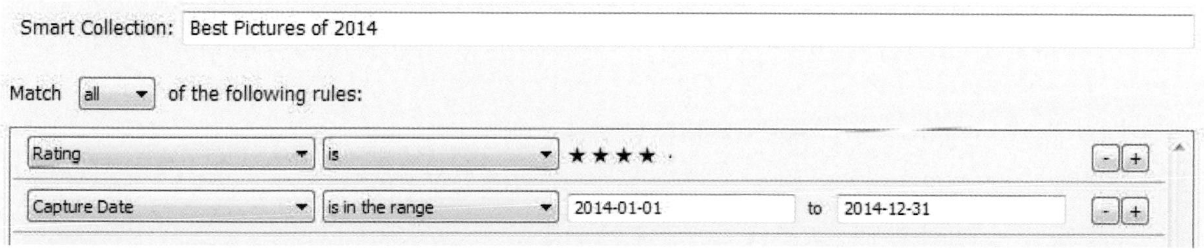

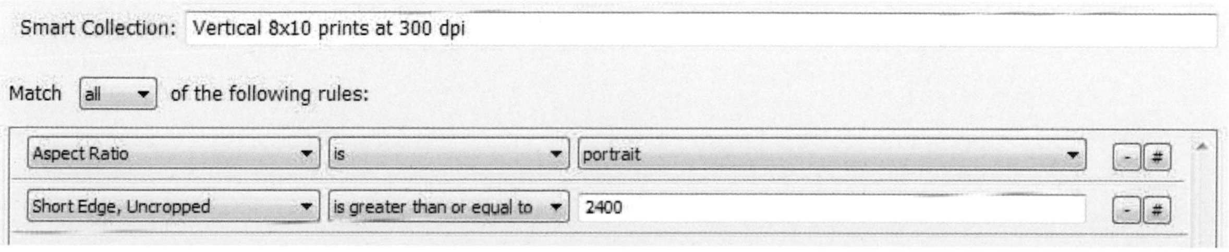

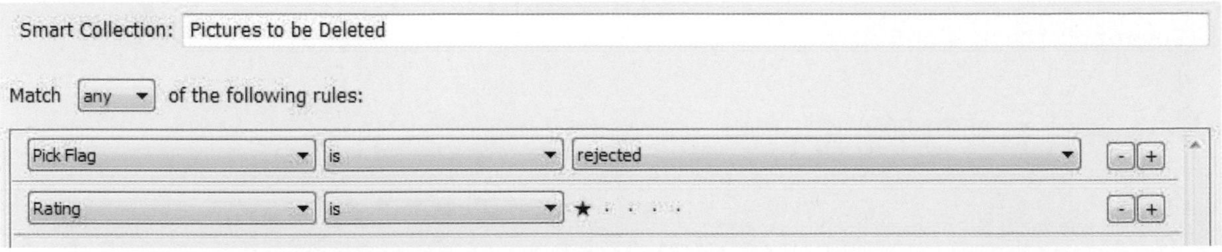

Smart Collection: Pictures to be Deleted

Match any ▼ of the following rules:

| Pick Flag ▼ | is ▼ | rejected ▼ | - + |
| Rating ▼ | is ▼ | ★ · · · · | - + |

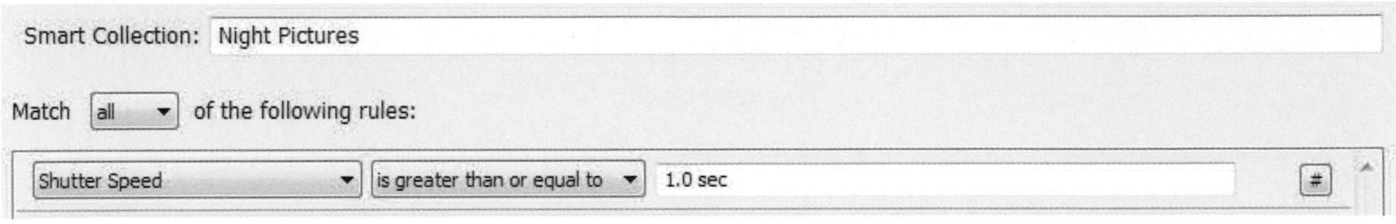

Smart Collection: Night Pictures

Match all ▼ of the following rules:

| Shutter Speed ▼ | is greater than or equal to ▼ | 1.0 sec | # |

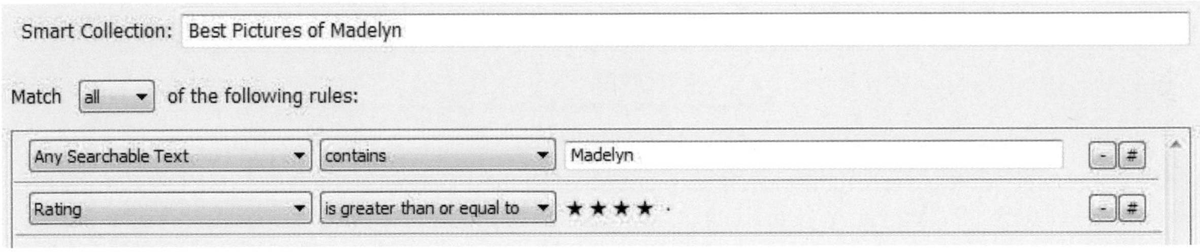

Smart Collection: Best Pictures of Madelyn

Match all ▼ of the following rules:

| Any Searchable Text ▼ | contains ▼ | Madelyn | - # |
| Rating ▼ | is greater than or equal to ▼ | ★ ★ ★ ★ · | - # |

Organizing Collections

You can drag-and-drop collections into collection sets by clicking a collection with the mouse, holding down the mouse button, dragging the collection to the collection set, and then releasing your mouse. You can also drag pictures into collections in the same way.

Using a Target Collection

You can specify one collection as the target collection to more easily add pictures to the collection. To specify the target collection, right click the collection and then select Set As Target Collection. You'll see a + symbol next to the collection name, as shown next. Now select a picture and press B on your keyboard to add the picture to the target collection.

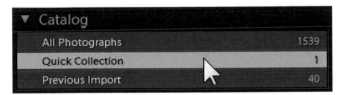

USING THE QUICK COLLECTION

The quick collection (which is, by default, also the target collection) is just a temporary spot to drop pictures into. If you just need to grab a handful of pictures, and you won't ever want to refer to them as a collection in the future, the quick collection is perfect. Adobe has hidden Quick Collection inside the Catalog pane, which makes absolutely no sense.

Because the Quick Collection is the default target collection, you can usually add pictures to it by pressing B on your keyboard. You can also right-click any picture and click Add To Quick Collection.

If you do put something valuable in your Quick Collection, you can right-click it and save it as a permanent collection. You can also clear the Quick Collection by right-clicking it.

FOLDERS

As I mentioned in Chapter 1, you shouldn't use Explorer or Finder to manage your pictures; you should use Lightroom instead. In the Library view, the Folders pane is underneath the Catalog pane. Within Folders, you can browse every drive attached to your computer, including external drives.

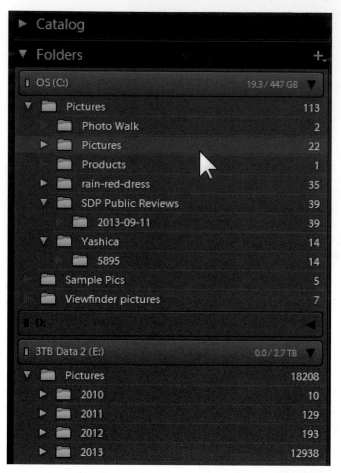

If you're using Lightroom correctly, you should almost never need to open the Folders pane. Instead, you should use Filters (as discussed in Chapter 1) and Collections to find your pictures. In fact, you can quickly create a collection from a folder simply by dragging the folder into the Collections pane.

The one time you do want to use the Folders pane is when moving pictures between drives. For example, if your C: drive is full and you want to move pictures to an external drive, you should use the Folders pane to drag the folder between the drives. Lightroom will move the files, just as Explorer or Finder would, but it will also update the location of every moved picture in your catalog.

If you want to add pictures to your library but you can't find the folder in the Folders view, add them to the current catalog by using the File | Import Photos And Videos menu item.

If you want to move pictures to a folder on another drive and that folder doesn't appear in the Folders pane, click the plus sign to open the Folders shortcut menu, and then select Add Folder.

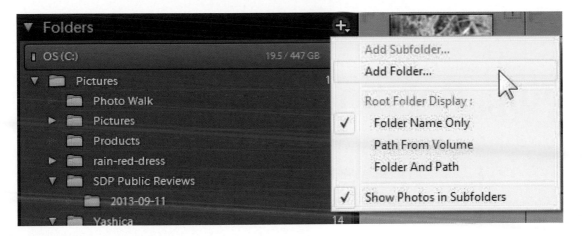

If you use Explorer or Finder (or any tool besides Lightroom) to move a folder in your Lightroom catalog, Lightroom will mark the folder as missing. Don't worry; just right-click the folder and select Find Missing Folder, as shown next.

SUMMARY

Collections provide a powerful and flexible way to organize your pictures in a hierarchy. While keywords, ratings, and labels are definitely useful, collections should be the primary tool you use to organize your pictures. If you want to be able to find a picture in the future, you should add it to at least one collection.

5
Cropping, Rotating, & Removing Spots

To view the videos that accompany this chapter, scan the QR code or follow the link:

SDP.io/LR5Ch5

When you leave the Library module and select the Develop module, you get access to a wide variety of very powerful tools. Many of the adjustments you learned in the Library module are still applicable—you still have the ability to adjust color temperature, brightness, contrast, and clarity. In the Develop module, however, you get much finer control. Not only can you dial in smaller increments of change, but you can apply changes to small portions of your picture, creating works of art that might otherwise require you to spend years studying Photoshop.

In this chapter, we'll start with the most basic tasks: cropping and rotating your picture and removing spots. Future chapters will cover more advanced aspects of the Develop module.

CROPPING AND ROTATING

Most pictures need a bit of a crop before they're presented. In fact, I recommend photographers shoot slightly wider-angle than they need to and leave a bit of room to crop in every picture. Leaving room to crop allows you to change the aspect ratio of an image without destroying your composition.

For example, if you take a family portrait and you want to print it at 8x10", you'll need to crop an inch of the sides of the photo. Then, when you mount the photo in a frame, the frame will cover even more of the edges. If you forgot to leave room to crop, the frame will be hiding half of someone's face.

ROTATING

Your first step in editing is to rotate your picture to level it. To rotate a picture, click the Crop tool or press **R**, and then hover your cursor near any corner of your picture. You'll see the cursor change to the rotate tool, as shown next. Now, hold your mouse button down while dragging the mouse to rotate the picture. Lightroom shows gridlines (by default) to help you level the horizon.

If there's a perfectly horizontal element in your picture, such as the horizon, or a perfectly vertical element, such as a building in the center of the frame, use the Level tool to rotate the image. After selecting the Crop tool (or pressing **R**) hold down the **Ctrl** key (on a PC) or **Cmd** key (on a Mac). The cursor will turn into a level. Then, drag your cursor from one end of the subject to the other. When you release the mouse button,

Lightroom will rotate your picture so that the line you drew is either perfectly horizontal or vertical. In the following example, I used the level tool to draw a line along the horizontal blade of the windmill, ensuring it was perfectly level in the photo.

Lightroom can try to automatically level your photos, too, and it's surprisingly accurate. In the Lens Corrections panel of the Develop module, click **Auto** and double-check that Lightroom doesn't mangle your picture. It's wise to select all the checkboxes on the Basic panel, too. We'll discuss the Lens Corrections panel in more detail later.

CROPPING

Rotating your picture always involves at least a bit of cropping, which is why I suggest rotating your image first. Once it's level, select the Crop tool or press **R**, and hover your cursor over a corner of the image until your cursor turns into the crop tool, as shown next. Now, drag your cursor in to crop your picture tighter.

If you want to crop deeply into a photo, select the Crop tool and then drag your cursor from corner to corner without grabbing the edges, as shown next.

Tip: Press **Shift+L** key while cropping to enter Lights Out mode and hide all distractions. Press **L** to return to your normal display.

CHANGING THE ASPECT RATIO

By default, Lightroom keeps your original aspect ratio when cropping. That's nice; if you share your pictures, they won't look cropped.

If you want your picture to be square, or you want to crop it to 8x10, click the **Aspect** list and select the preferred ratio. If you want to switch your image from horizontal to vertical, or vice-versa, drag any corner until Lightroom switches the orientation of the crop.

If you don't want to be constrained to a pre-defined aspect ratio, click the Lock icon to unlock it, as shown next.

USING LOUPE OVERLAYS

When you crop a picture, Lightroom overlays a rule of thirds grid on your picture to help you follow that compositional guideline more precisely. When you rotate a picture, Lightroom overlays a tighter grid to help you level your horizon and straighten buildings.

Those overlays are enough for most photographers. If they're not enough for you, you can add Loupe Overlays (**View** | **Loupe Overlay**) in either the Library or Develop module. Lightroom provides three types of loupe overlays: grids, guides, and layout images, described in the next sections.

Tip: Quickly turn any loupe overlay on or off by pressing **Ctrl+Alt+O** (on a PC) or **Cmd+Opt+O** (on a Mac).

Grids

To display a grid over your picture, view the toolbar (if necessary) by pressing **T** or selecting **View** | **Show Toolbar**. Then, click the triangle on the right side of the toolbar and select **Grid Overlay**.

Lightroom adds the Grid Overlay tool to the toolbar. By default, it'll automatically show a grid when you drag the slider. If you want to see the grid or adjust the opacity of it, click the list and then select **Always**, shown next.

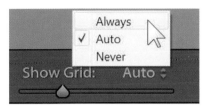

When the Grid is set to **Always**, you can hold the **Ctrl** key (on a PC) or the **Cmd** key (on a Mac) to view the grid controls, shown next. Drag the **Size** and **Opacity** numbers to adjust your grid.

Guides

Guides draws a horizontal and vertical line on your picture wherever you want it. It's a great way to check the alignment of horizontal or vertical subjects. From the menu, select **View | Loupe Overlay | Guides**. To move the guide, hold down the **Ctrl** key (on a PC) or the **Cmd** key (on a Mac), as shown next.

Layout Images

Lightroom can overlay your picture with a partially transparent image, which is useful for photographers cropping pictures for pamphlets, books, and magazines.

Here's a bit of insight into the printing world: anytime you see a picture in printed material that goes right to the edge of the paper (like the cover of this book), the original image actually went beyond the edge of the paper. Cutting paper isn't an exact process, so designers use a template with a trim line around the outside that might or might not get cut. If an image should bleed to the edge of the page, it must go past the trim area. There's also a margin area where text shouldn't fall, or it would crowd the edge of the paper.

When designers choose to overlay text on an image, they require that the image has the same aspect ratio as the page, and that the image has enough negative space for their title and text. If you're working with a designer, they can probably provide a template for you, or you can make a template in Photoshop.

Now, add the PNG file to Lightroom. From the menu, select **View | Loupe Overlays | Choose Layout Image**. Crop your picture to fit your layout.

The next example shows a two-page magazine layout. You can download the template at *sdp.io/ch5*. The green areas are safe for text and images, the yellow areas are the margins, and the red areas are the bleed that might or might not be cropped. By overlaying the layout on an image with negative space, I can verify that the text doesn't cover important parts of the picture, and that the subject of the photo won't be cropped.

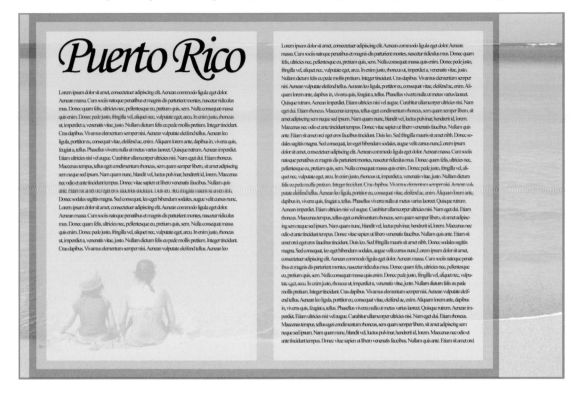

Hide or show the overlay by selecting **View | Loupe Overlays | Layout Image** or by pressing **Ctrl+Alt+O** (on a PC) or **Cmd+Opt+O** (on a Mac).

You can adjust the size, placement, opacity, and matte of the layout image by holding down the **Ctrl** key (on a PC) or the **Cmd** key (on a Mac). The opacity is how transparent the layout image is, and the matte is how much Lightroom darkens parts of your image that are outside the layout.

However, Lightroom has a bug that won't allow you to adjust the **Opacity** or **Matte** by default. When you try to drag the numbers on the control bar, Lightroom shows the hand cursor to drag the layout to a different place, as shown next.

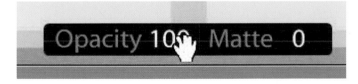

To work around this bug, hold down the **Ctrl** or **Cmd** key, drag the layout out of the way, and then adjust the **Opacity** or **Matte** values. Then, drag the layout back to its original position.

REMOVING SPOTS

While it's not as powerful as Photoshop, Lightroom's Spot Removal tool can do a good job of removing dust and distractions from your images.

The Spot Removal tool has two modes: Clone and Heal. Heal is the right choice for most situations; it'll intelligently blend another part of the picture over the spot that you click.

Heal works remarkably well, but you will occasionally need to use the Clone mode. Clone copies another section of your picture over a spot, without using any intelligence whatsoever. If Heal fails, you might be able to find a part of the picture to clone.

Size is the Spot Removal tool's most important property. You should adjust your size so that your brush is just slightly larger than the spot you want to remove. You can use the Size slider, but it's much faster just to scroll your mouse wheel.

With the size set, simply click any spot you want to remove. In the next example, I'm removing a spot of dust from the sky.

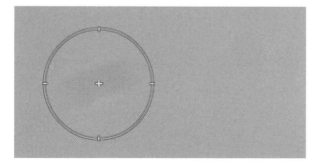

Spot removal can also remove lines and curves; simply drag your cursor along the entire object to be removed.

MANUALLY SETTING THE HEAL SOURCE

Occasionally, Lightroom will use a completely stupid part of the picture to heal the spot you want removed, as the next examples show. Assuming you don't want a nose floating in the middle of the sky, simply drag the circle indicating the Spot Removal's source to a part of the picture that matches the area you want to heal.

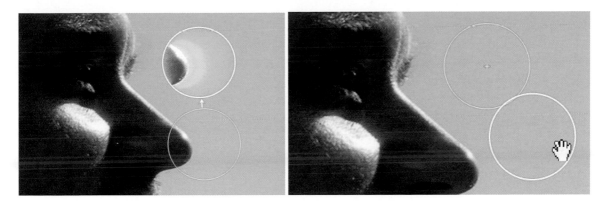

If you use Spot Removal to heal dust that appears in multiple pictures, you can quickly apply those spot removal settings to multiple pictures by copying and pasting the settings, as described in Chapter 1. Beware, however, that Lightroom will try to use the same part of every picture as the source. Therefore, you should carefully check the spot removal in every picture.

ADJUSTING FEATHERING

If you use the Clone brush, you might need to adjust the Feathering property. Feathering smoothly blends the cloned spot into the surrounding picture, giving a more natural appearance. For example, this example shows cloning with **Feathering** set to **0**. As you can see, the edges of the spot being replaced are sharp, and the cloning looks completely unnatural.

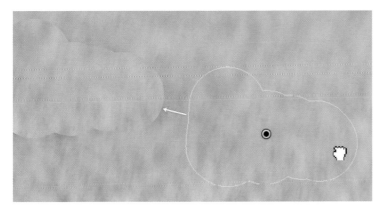

The next example shows the same cloning with the **Feathering** set to **94**. As you can see, the sharp edges are gone, and the cloning looks much more natural.

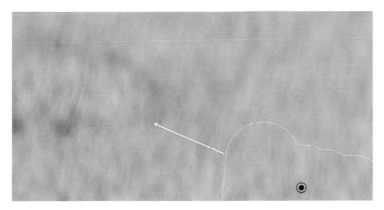

ADJUSTING OPACITY

Opacity allows the spot you're covering to show through. With **Opacity** set to **100** (the default) your cloned spot completely hides the background. With **Opacity** set to **50**, the cloned area and the background will be blended evenly, as shown in the following example, in which you can partially see the shadow I was attempting to erase from the grass.

You don't usually need to adjust the Opacity, but it can be useful for making cloning more subtle.

REMOVING SPOT CORRECTIONS

If you want to get rid of a spot removal, select the spot removal tool and hold down the **Alt** or **Opt** key on your keyboard. Your cursor will turn into scissors. Click any spot to remove the correction.

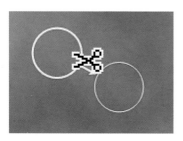

USING RED EYE CORRECTION

Red Eye can occur when you use a flash too close to the lens. Basically, the back of the eye is reflecting only the red frequencies of light. If you take pictures of cats with a flash, you'll find it much easier to create red eye (or often green eye) because of their larger pupils.

The best way to fix red eye is to prevent it when taking a picture. Instead of using on-camera flash, use an external flash. Better yet, bounce your flash off the ceiling. For detailed information, refer to Chapter 3 of *Stunning Digital Photography*.

If you're forced to use on-camera flash, Lightroom can help you quickly remove red eye. Select the Red Eye Correction tool and zoom in on your subject's eyes.

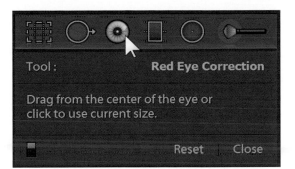

Now, you just need to select the eyes. Use your mouse's scroll wheel to make the circle about the same size as the eyes, and then click them both. If that doesn't work, click in the center of the red eye, and then drag outward until the red eye is completely covered. If Lightroom covers only part of the red eye, you can grab the edge of the circle and drag it outward.

The next figure shows Chelsea with red eye fixed in only one eye.

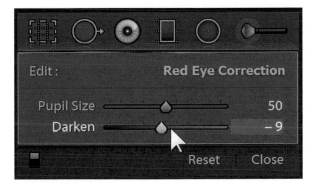

You can also use the **Pupil Size** and **Darken** sliders to tweak the effect to your liking.

SUMMARY

The tools at the top of the Develop module are some of the most powerful and complex. They are capable of creating amazingly complex edits, rescuing pictures that most photographers would have thought were simply ruined.

For best results, be subtle. Make the changes you think make your picture look best, and then dial them back by 50%. Most people prefer a more natural picture, but it's easy to overdo it when you're editing.

6
Filters & Adjustment Brushes

To view the videos that accompany this
chapter, scan the QR code or follow the link:

SDP.io/LR5Ch6

In this chapter, we'll cover some of Lightroom's most powerful tools: filters and adjustment brushes. These tools allow you to selectively apply changes to parts of your picture. I'll show you how to restore the sunset to an overexposed sky and how to smooth skin to diminish the appearance of pores and blemishes in a portrait.

USING GRADUATED FILTERS

Graduated filters allow you to adjust part of your picture without changing other parts. For example, you could darken an overly bright sky, without changing the foreground. When you select the graduated filter tool, as shown next, Lightroom gives you a variety of slides that you can use to control the changes it makes to your picture.

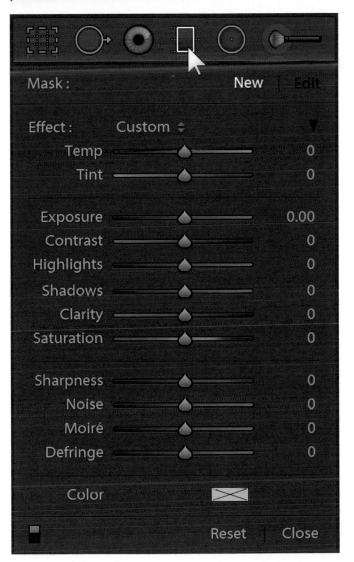

Graduated filters fade changes gradually across your frame, using a straight line to divide the changed and unchanged parts of the picture. This means they work well for landscapes, but don't work so well when a shape isn't straight, such as when changing the sky behind a person or animal. For more complex shapes, use the Adjustment Brush.

Apply the graduated filter by dragging your cursor from the side where you want to make the most change to the side where you want to make the least change. The transition is smooth and feathered, and you'll usually get more natural results if you drag the filter a little past where you want to change your picture.

In this next example, I want to reduce the brightness of the sky without darkening the foreground. The picture is properly exposed, but the sky seems to be solid white. To my eye, the sky had beautiful blue colors with clouds lit by the setting sun. However, the limited dynamic range of my camera hid all those details.

Fortunately, I can fix this quickly with Lightroom. I selected the graduated filter tool, lowered the **Exposure** and **Highlights** sliders, upped the **Clarity** and **Saturation** a bit, and then dragged from the top to the bottom of the transition period between the sky and the foreground. Everything above the graduated filter has the full effect of my settings, with no feathering. Everything below the graduated filter is completely unchanged. In the area of the graduated filter, Lightroom smoothly blended the effect so it looks natural.

You can always adjust the settings after you apply the graduated filter. Grab the edge of the filter with your mouse to make it bigger or smaller. Grab the center of the filter (with the black dot) to move the filter around the image.

Now, the Stockholm sky looks much more like what my eyes saw. However, the foreground is still a bit dark. I can use another graduated filter to raise the exposure of the foreground without changing my sky. This time, I dragged from the bottom towards the top, so my changes would impact the bottom half of the picture.

Tip: If you drag the graduated filter in the wrong direction, press the ' (apostrophe) key to flip it.

Using Radial Filters

There's still one problem with the photo—the balloon is now dark. When I saw it with my eyes, it was lit by fire, and nice and bright. The graduated filter I applied to the sky darkened it unnaturally.

I can't use a graduated filter to fix the balloon, however, because it would cover the entire width of the picture. However, the balloon is a nice round shape, so I could use a radial filter. Like a graduated filter, the radial filter smoothly blends your changes.

First, select the radial filter tool.

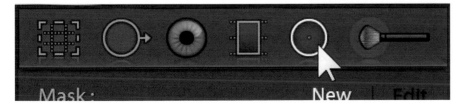

Now, drag from the outside of the circle towards the center, until the part of the picture you want to adjust is highlighted. If you drag from the center outward, just select the **Invert Mask** checkbox at the bottom of the panel, as shown in the next example. If you drag diagonally, you'll get a circle. If you drag up, down, or to the side, you'll get an oval.

As you can see, I added some exposure to return the balloon to its original brightness. The radial filter gave it a nice glow.

USING THE ADJUSTMENT BRUSH

Graduated filters are great for making adjustments along a line, and radial filters are perfect for circular or oval shapes. Most of the earth isn't simply straight or round, however. If you want to make filters with more complex shapes, use the adjustment brush, as shown next.

Here's a picture Chelsea snapped of me in front of the same sunset. Looking at the picture, however, you wouldn't even know there was a sunset behind me—it's completely overexposed.

To recover the sky, I selected the adjustment brush and used my mouse wheel to make the brush very large, and then painted the sky around my head. I adjusted the Feather size up, and overlapped the feathered area of the brush across the subject. To allow Lightroom to intelligently find the edges, I selected the Auto Mask checkbox. As you can see, the sky reappears, showing a scene much more like what we saw with our eyes.

Use a large brush and a high feather size to make smooth, even edges. To fine-tune your edges, use a small brush and lower feather sizes. If you make a mistake applying your mask, hold down the **Alt** key (on a PC) or the **Opt** key (on a Mac) and paint over the mistake. **Alt** or **Opt** erases parts of your mask.

Looking at the picture, I see that my jacket is entirely black. I'd like to show some more of the texture of the jacket without raising the shadows in the entire picture. To create a new adjustment brush, click **New**.

Next, I painted a new mask over my jacket and raised the Exposure, as shown next. You can definitely see the detail in my jacket, but it looks awful.

When I raised the exposure, Lightroom showed a great deal of noise in my jacket. That's a common problem when raising shadows. To offset that, I raised the **Noise** and **Moiré** sliders to the maximum and lowered the Saturation slider to eliminate any unnatural colors. Now, you can see the detail the detail in my jacket without any noise.

Tip: Hold down the **Shift** key to paint in a straight line.

USING THE ADJUSTMENT BRUSH TO ADD LIFE TO AN EYE

For wildlife photography, the adjustment brush is very useful for improving the light in an animal's eyes. For example, while the bluejay in the next example does have a catch light from the sun, the eye is almost completely black.

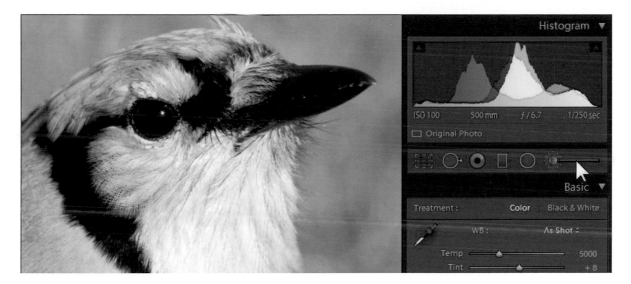

I could raise the shadows in the entire picture, but the adjustment brush lets me make changes to just the bird's eye. I made a brush slightly larger than the bird's eye and adjusted the exposure and shadows up to show the detail in the bird's iris. This adds life and personality to the eye.

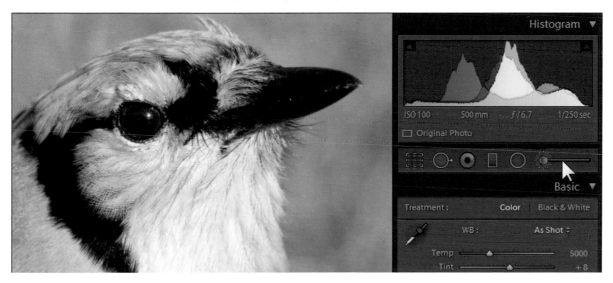

If you like the idea of the adjustment brush, but you hate the rough edges, or you wish you had more control, I suggest using layers and masking in Photoshop. Photoshop's masking capabilities are far more advanced.

USING THE ADJUSTMENT BRUSH TO ADD A CATCH LIGHT

While I prefer the more subtle approach shown in the previous example, sometimes an eye is so dark that there's absolutely no natural light in the eyes to brighten. It's always better to get a natural catchlight; keep the sun to your back, or use a bit of fill flash. If that's not possible, you can use the adjustment brush to add a fake catch light in post-processing.

Consider this Cedar Waxwing, which has an almost completely black eye.

To add a fake catchlight to his eye, I made the smallest possible adjustment brush and disabled **Auto Mask**. I then clicked the upper corner of his eye (where you might see a natural catchlight from the sun) and raised the **Exposure**, **Highlights**, and **Shadows**.

My settings are shown here:

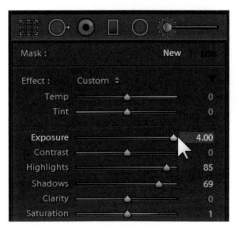

USING THE ADJUSTMENT BRUSH TO SMOOTH SKIN

Today's cameras and lenses are often too sharp for portraiture; you can see every pore and blemish. You might not even notice them during a conversation with the person because you'll be making eye contact. When you're staring at a still photo, however, your eyes roam to the texture of their skin.

Except for those of young children, every portrait needs a bit of processing on the skin. Photoshop's masking tools are clearly superior for this, but you can quickly smooth skin in Lightroom using an adjustment brush and the Clarity slider.

First, the unedited portrait of the author, by Chelsea.

Now, I'll select the **Show Selected Mask Overlay** checkbox to see where I'm painting. Then, I'll select the **Auto Mask** checkbox, add feathering to the brush, and adjust the **Clarity** slider to -50 (or so). This example shows that the adjustment brush is applied to my face; the mask overlay makes the selected areas appear red:

Turning off **Show Selected Mask Overlay** reveals the results (and allows you to fine-tune the **Clarity** slider). The skin looks much smoother than in the unedited photo. If you want to make a portrait grittier, perhaps by accentuating freckles or wrinkles, use this technique and drag the **Clarity** slider to the right. Don't drag the **Clarity** slider to the right on a portrait of yourself unless you're prepared to make a teary-eyed call to the dermatologist.

USING THE ADJUSTMENT BRUSH TO ADD MAKEUP

You can also use the Adjustment Brush to add makeup. The model in the next example wore very natural makeup. First, the before picture:

Create a new Adjustment Brush. If you're applying lipstick, select the **Auto Mask** and **Show Selected Mask Overlay** checkboxes to select the lips, as shown in the next example.

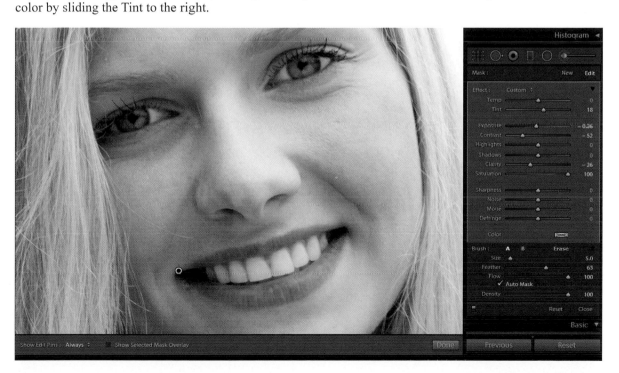

Her lips already look more red, but that's just because of the mask overlay. Once you're confident in your mask selection, clear the checkbox. Next, adjust the saturation up to bring out the natural colors. You might also consider dropping the **Contrast**, **Clarity**, and **Exposure**, as shown in the next example. You can add color by sliding the Tint to the right.

You can use the Color Picker to add color. This next example shows adding blue eye shadow. The higher the color you select from the color picker, the more intense it will be. To add subtle color, choose a color closer to the bottom of the color picker.

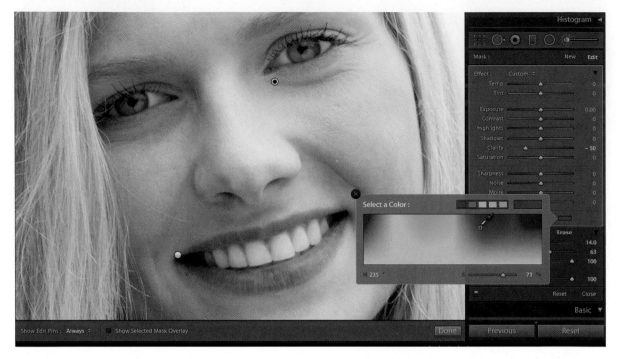

When adding eye shadow, it's easiest to select the entire eye with a large brush. Then, hold down the **Alt** key (on a PC) or the **Opt** key (on a Mac) and use a smaller brush to remove the mask from the eye and surrounding areas.

Using the Adjustment Brush to Whiten Teeth

Another edit you should make to most portraits is to whiten teeth. Even if someone has near-perfect teeth, as with this model, we're accustomed to seeing overly perfect teeth in magazines and on TV, so natural teeth actually look unnatural.

Use the **Auto Mask** to select the teeth. Then, adjust the **Exposure** up to brighten the teeth and the **Saturation** down to reduce any yellowing. Remember, the best changes are very subtle.

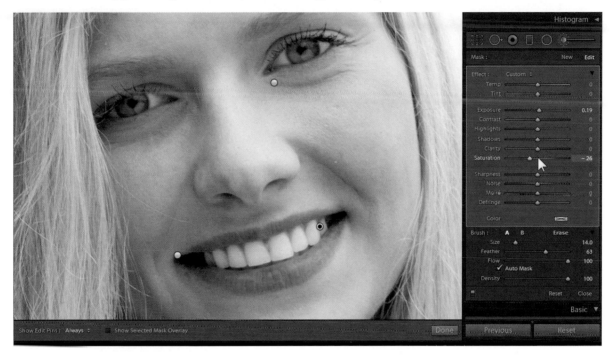

Hiding the Pins

The adjustment brush can be addicting. Once you add a few adjustments, the pins can be really distracting. To hide them, use the **Show Edit Pins** option in the bottom-left of the screen and set it to **Never**. For everyday use, the **Auto** setting (shown next) works better than the default of **Always**.

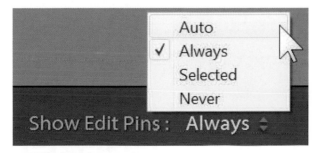

Summary

Lightroom does provide quick-and-easy tools for selectively adjusting parts of your picture without impacting the entire picture. They're also good enough to meet the needs of most photographers.

If you need more perfect control over your images, consider using Adobe Photoshop. Photoshop supports layers and masks, and offers a variety of plugins that perfectionists will require.

7
The Tone Curve Panel

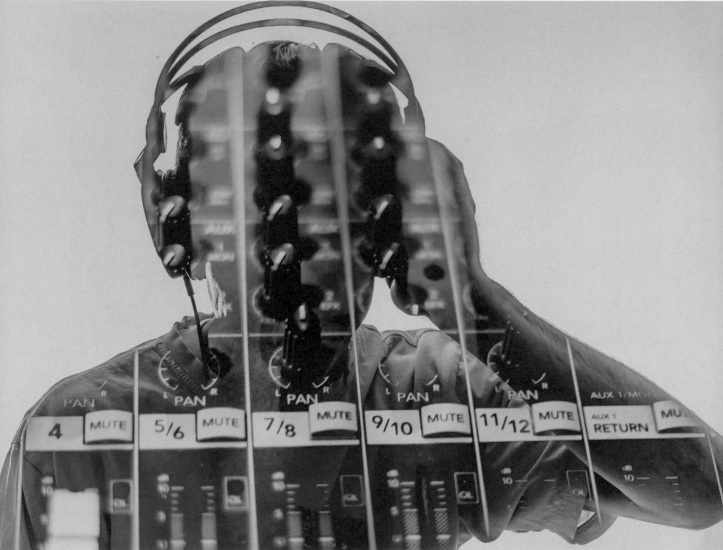

To view the videos that accompany this chapter, scan the QR code or follow the link:

SDP.io/LR5Ch7

This chapter describes how to use the Tone Curve panel in the Develop module to fine-tune the brightness and contrast of your photo. You'll also learn how you can separately edit the red, blue, and green channels hidden within your photo.

ADJUSTING THE TONE CURVE

The Tone Curve panel provides an alternate way to control your photo's overall brightness or the brightness of the shadows, mid-tones, and highlights. While you can accomplish most adjustments using only the Quick Develop or Basic panels, the Tone Curve provides finer control and a visual representation that might be easier for some to understand.

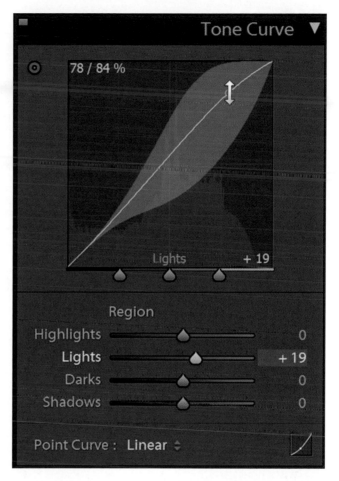

The Tone Curve panel is small, but incredibly complex. Many photographers will never need to use it; if you're feeling overwhelmed by Lightroom's technical aspects, skip this section and come back to it if you ever feel like you don't have enough control over your picture's brightness.

UNDERSTANDING THE GRAPH

The Tone Curve's graph has three elements:

- **A histogram.** Shown faintly behind the graph, this shows the distribution of dark and light in your image. You can't make direct changes to it; it's just for reference.

- **The tone curve.** The diagonal line shows how Lightroom maps dark and light parts of the original image to what it displays. Later examples will make this more clear.

- **Region dividers.** These markers at the bottom of the chart define the brightness that is considered shadows, darks, lights, and highlights. You could move these to make everything brighter than middle grey considered a highlight.

You can directly grab the tone curve to adjust shadows, darks, lights, and highlights:

- **Shadows.** The left-most part of the histogram, these are the parts of the picture that are almost black.

- **Darks.** Located between the middle marker on the graph and the shadows, these are the darker parts of the picture.

- **Lights.** Located between the middle marker and the highlights, these are the brighter (but not the brightest) parts of the picture.

- **Highlights.** The right-most part of the histogram, these are the parts of the pictures that are almost white.

For example, the next three examples show an unedited picture, and then the same picture with the Highlights (the right-most part of the tone curve) dragged all the way up and then down.

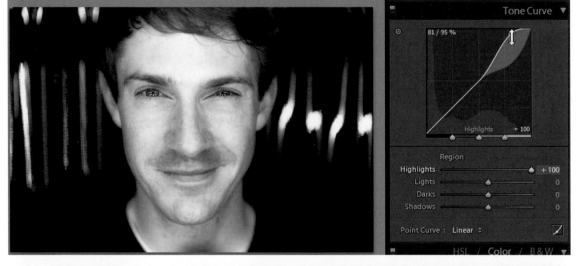

Because I adjusted only the highlights in the tone curve, the darker parts of the picture were completely unchanged. Next, I'll adjust the dark parts of the picture, without changing the highlights.

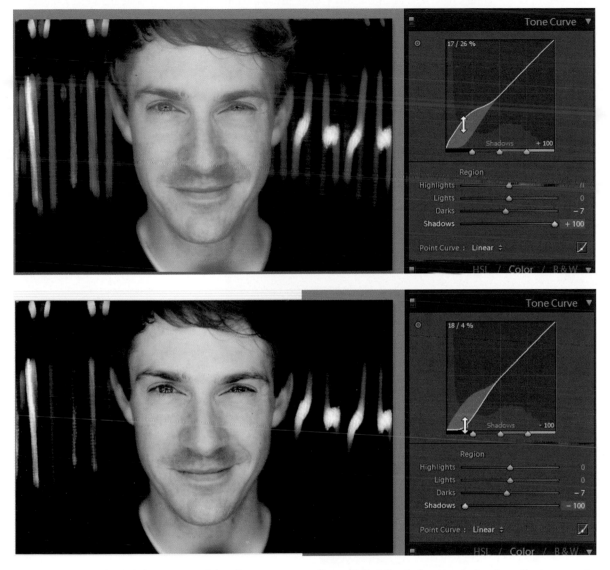

As you can see, the brighter parts of the picture were unchanged, but the darkest parts (such as those around the model's hair) were much brighter or darker.

Notice that wherever you drag it, the tone curve is always smooth. If it were to become jagged, so would your picture. This helps create a more natural look, but it's still entirely possible to overdo it by adjusting the tone curve too severely.

I made these changes by dragging the tone curve line on the graph. However, you could accomplish the exact same thing by moving the sliders below the tone curve. When you adjust the line, Lightroom also adjusts the values displayed by the sliders; they're simply two different user interfaces for one adjustment.

ADJUSTING THE CONTRAST

At the bottom of the Tone Curve panel is the **Point Curve** list, which allows you to choose between **Linear**, **Medium Contrast**, and **Strong Contrast**. **Linear** is the default—and the most natural—setting. If you choose **Medium Contrast**, some brighter midtones will be pushed into the highlights, and some darker midtones will be pushed into the shadows. If you choose **Strong Contrast**, even more of the midtones will be pushed into the brightest and darkest parts of the picture.

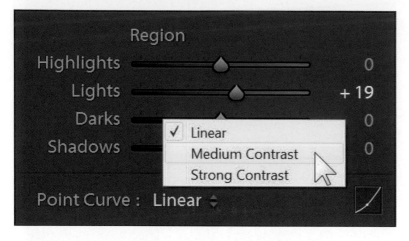

ADJUSTING THE SPLITS

By default, Lightroom divides your picture's brightness into four quarters, each taking 25% of the total available brightness: Shadows, Darks, Lights, and Highlights. You can adjust the split between each region by dragging the triangles at the bottom of the graph. This gives you more flexibility to control the brightness of specific regions of your picture. The following example shows how you might adjust the regions if you needed fine control over just the brightest parts of your picture:

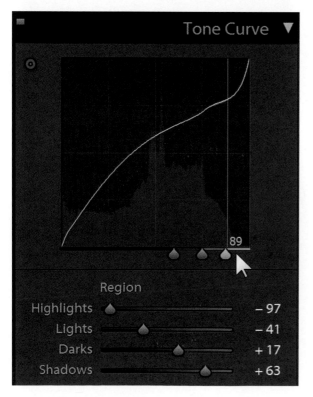

CREATING A CUSTOM POINT CURVE

As I mentioned, Lightroom tries to keep your tone curve fairly natural. Dark parts of the picture should be dark, and bright parts should be bright—the most you can do is push each part of your histogram to the left or right a bit.

If you want to get really creative, you can use a custom Point Curve by clicking the easy-to-overlook button in the lower-right corner of the Point Curve panel, as shown by the cursor in the next example. This allows you to create extreme effects, such as making the mid-tones the brightest part of the picture, as shown next.

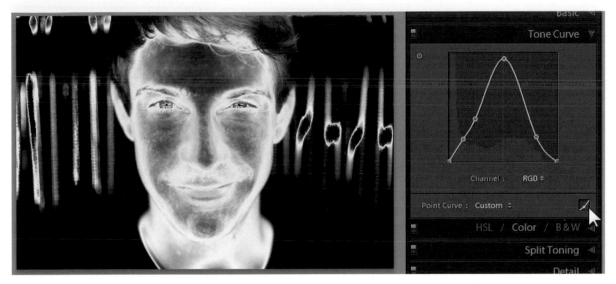

You can add as many points as you want to with a custom point curve; simply click the tone curve and drag it where you need it. You can use this to make a specific range of tones brighter or darker than the surrounding tones, as shown in the next example.

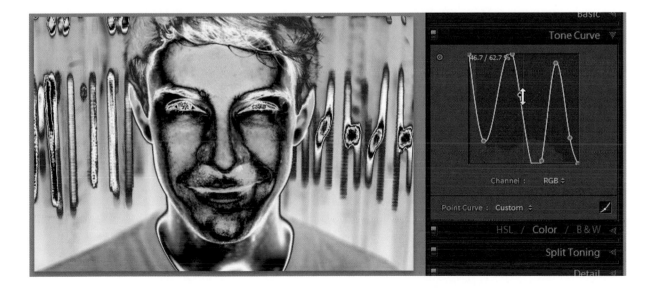

Besides creating extreme portraits, the technique has practical applications, especially for landscape photographs. In this picture, the sky seems completely white in the photo. However, that's not the way the sky looked to my eyes—the clouds were big and fluffy, and the sky had a distinct texture.

To improve this, I added many points to a custom tone curve and then adjusted the brightest highlights so that the tone curve fell off sharply. Whereas the entire sky in the original photo was 99-100% bright (and thus you couldn't see the texture) with the tone curve, the sky was now 90-100% bright, allowing you to see the texture in the sky. I also brightened the mid-tones of the church steeple to make it brighter than it had originally been, so it wasn't overpowered by the sky.

DIRECTLY ADJUSTING PARTS OF YOUR PICTURE

Most people don't look at a picture and divide it into shadows, darks, lights, and highlights. Instead, we look at a landscape and think, "I wish I could make that sky darker," or "I wish there were more contrast in those hazy hills."

Lightroom lets you adjust the tone curve by directly grabbing different parts of your picture. Click the icon in the upper-left corner of the Tone Curve panel, as shown next. Then, grab a part of your picture and drag it up or down. Lightroom determines whether you're grabbing a shadow, dark, light, or highlight, and adjusts it based on your mouse movements.

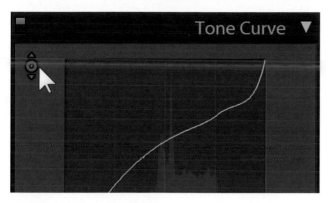

You can use this feature with custom point curves. However, you must add points to the curve before you can drag the brightness of a section up or down. I suggest clicking the point curve in several different places, as shown next, before dragging parts of your picture.

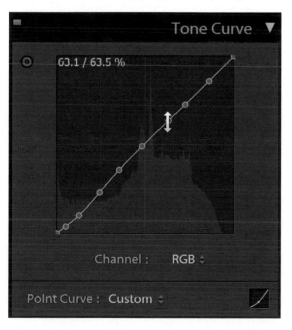

ADJUSTING INDIVIDUAL CHANNELS

Lightroom allows you to directly adjust the red, blue, and green channels in a picture. Here's a quick biology and physics lesson, just in case this doesn't make perfect sense.

UNDERSTANDING RED, GREEN, AND BLUE (RGB) CHANNELS

The human eye has three different types of color receptors: red, green, and blue (unless you're color blind). Those are the only three colors eyes can distinguish; other colors are just combinations of the three primary colors. Red and blue make purple; red and green make yellow.

Because most humans can only see three colors, pictures only need to be three colors to recreate the full visible spectrum. Therefore, most cameras only capture red, green, and blue light; yellow light frequencies are caught in both green and red pixels, and purple light frequencies are caught in both red and blue pixels. Similarly, your monitor has separate red, green, and blue pixels, and if you get close enough, you can see them.

Pictures are (basically) stored as three channels within your picture: a red channel, a green channel, and a blue channel. The next example shows a picture of a House Finch and its red, green, and blue channels.

Remember, yellow is made of both red and green, so the red and green channels show the background as being very bright. Notice that the left part of the background has less green than the right part; in the original picture, the left part of the background is more orange (which has more red than green) than yellow (which has equal amounts of red and green).

The blue channel shows the background as being almost black, because blue isn't part of the mix that represents yellow. In fact, the picture has no blues, purples, or turquoise at all, so you might think the blue channel would be completely black. However, red, blue, and green all combine together to create white or grey. Therefore, the blue channel shows the parts of the original scene that our eyes would see as shades of white.

ADJUSTING RGB CHANNELS

You can use the Tone Curve panel to separately adjust the red, green, and blue channels within your pictures to correct color problems or to add special effects. For parts of your picture that consist of one of the three primary colors, such as a red shirt, a blue sky, or green grass, this gives you fine control over the brightness of the highlights and shadows for just that part of the picture.

To adjust individual channels, click the point curve icon in the lower-right corner of the Tone Curve panel, as shown next.

As the next example shows, you can select which of the red, blue, and green channels you want to adjust.

The next example shows a before and after of our House Finch with the blue channel raised. Because the blue channel is now brighter than the red and green channels on the (formerly) white and grey parts of the bird, parts of the bird have shifted towards purple.

If we remove the blue entirely, we see that white and grey parts of the picture become yellow. Remember, whites are made of a mix of red, green, and blue. Removing blue, we now have a mix of only red and green, and red and green make yellow.

CREATIVE TONING WITH RGB CHANNELS

You can adjust individual channels to tone a picture, changing the mix of red, blue, and green within the shadows, mid-tones, and highlights of a picture. For best results, use very gentle curves.

In this first before-and-after example, I used two control points to put a slight curve on the reds, raising the effect of the reds in the highlights, but dropping them in the shadows. By raising the highlights, Chelsea's skin tones became pinker. By lowering the shadows, the darker parts of the picture (the smoke and Chelsea's hair) became more blue/green.

Switching to the green curve, I used a control point to keep the shadows at their current level. I used a second control point to slightly raise the highlights, shifting the pink skin tones more towards yellow.

Finally, I'll add two control points to the blue tone curve, bringing up the shadows and bringing down the highlights. This will move highlights towards yellow, since the red and green channels will now overpower the blues. However, the blues will overpower the reds in the shadows.

Though you don't have as much control, it's usually easier to use Split Toning to accomplish these types of effects, as described in Chapter 8.

SUMMARY

You won't often have to use the Tone Curve panel; the vast majority of brightness and contrast problems can be solved using the Quick Develop panel in the Library module or the Basic panel in the Develop module.

However, the Tone Curve panel can create much more drastic changes. You can make severe adjustments to specific ranges of shadows, mid-tones, or highlights. You can even make shadows brighter than highlights, if you have a reason to.

8
Color, Split Toning, & Effects

To view the videos that accompany this chapter, scan the QR code or follow the link:

SDP.io/LR5Ch8

In Chapter 1, you learned to use the Vibrance and Saturation tools to make adjustments to the entire picture. In this chapter, we'll cover the HSL / Color panel, and how you can use it to change the brightness (also known as luminance), hue, or saturation of specific colors. For example, you could make yellow grass green, darken a blue sky, or reduce the distracting saturation of a child's bright purple shirt.

To the beginning photographer, these tools are useful for problems such as incorrect white balance or an underexposed picture. More advanced photographers can use these tools to create film-like toning that supports the mood of the original photo or better blends into a set of photos.

This chapter describes the HSL / Color / B&W and Split Toning panels. Converting pictures to black-and-white is covered in Chapter 9.

ADJUSTING LUMINANCE

How many times have you taken an outdoor picture on a beautiful day, only to have the sky look dull and white? With just a couple of clicks, you can easily restore the brilliant blue you saw.

In the HSL / Color / B&W panel, click **HSL**. Select the **Luminance** tab, and click the direct adjustment tool, as shown next:

With the direct adjustment tool selected, click the sky (or whichever color you want to adjust) and drag the cursor up or down. For the following example, I dragged the **Luminance** down, making the blue sky darker. This also gives the appearance of a more saturated blue. Usually, when you want the sky to have more pop, you want to adjust luminance rather than saturation.

In the previous example, I dropped the blue **Luminance** to -48. That's a more extreme change than I would ever recommend; I just wanted to be sure the changes would be obvious for the sake of the example. Indeed, one of the most common problems I see in reader pictures is that they over-adjust the sky. Small tweaks are almost always better.

Notice that I used the direct adjustment tool to grab part of the picture, rather than simply dragging the **Blue** slider to the left. That technique would work just as well, but I often misjudge aqua as blue, or magenta as purple, requiring some trial and error. If you use the direct adjustment tool, you won't have to worry about guessing the right color.

ADJUSTING SATURATION

Luminance adjusts the brightness of a single color, while saturation adjusts the richness. For the sake of comparison, here's the previous picture with the blue saturation raised +48. The brightness is the same as the original picture, but the colors themselves are richer. Often, I'll raise the saturation of the sky a little (say, +10) and lower the luminance a bit (perhaps -10) to maximize the effect while maintaining an almost-natural appearance.

In the next example, my daughter is examining a massive caterpillar. To my eye, the reds in my daughter's skin tones are too saturated—probably because she was wearing a red shirt that was reflecting into her face.

To fix this, I clicked the **Saturation** direct adjustment tool, clicked her face, and dragged the mouse downward until I was happy with the result. As you can see, Lightroom found both orange and red tones in her face, and adjusted them both (relative to the mix of orange and red in her skin tones) to provide the most natural look possible. The green in the caterpillar is untouched.

WHAT IF TOO MUCH CHANGES?

If you make a saturation or luminance adjustment, you might discover that the color appears in parts of your picture that you don't want to adjust. Instead of using the HSL panel, use the adjustment brush to select just the part of the picture that you want to change, as described in Chapter 6. Then, adjust the saturation or vibrance of the single object.

LUMINANCE, SATURATION, AND VISUAL WEIGHT

As discussed in Chapter 2 of *Stunning Digital Photography*, it's important that the subject of your picture has the most visual weight. An object's visual weight consists of many different factors, including size and contrast, but two of the most important factors are luminance and saturation.

Basically, the brighter an object is, and the richer its color, the greater its visual weight, and thus the more prominent it appears in your picture. If an object is brighter or richer than your subject, it might be distracting, as the following example shows.

For example, in the following snapshot, the subject is my daughter and the waterfall. However, her pink sweater is very distracting because of both the color and luminance. By comparison, her blue jeans are not a distraction, because the luminance is much lower, and the color isn't as rich.

To improve this, I used the direct control slider to adjust the luminance and saturation of the pink color in her sweater. Though I tried to keep the changes subtle, the sweater is now much less of a distraction. Now, the viewer's eyes first go to my daughter's face and then to the waterfall, instead of simply to her sweater.

Thus, you can use luminance and saturation to improve the composition of your photo by reducing the visual weight of distractions.

SPOT COLOR/SELECTIVE COLOR

Spot color (also known as selective color) is a technique that removes all color from a photo except for the color in one specific subject. Basically, it's a black-and-white photo with one color object.

You can use the Saturation panel to create spot color. First, drag all the sliders to the left, as shown next.

Now that you have a black-and-white picture, use the direct adjustment tool to raise the saturation of your subject back to 0.

ADJUSTING HUE

Hue describes whether a color is red, orange, yellow, blue, green, or purple. You'll use it less than saturation and luminance, but it's great for selectively correcting or fixing colors.

Here's a trick that's cheaper than hiring a landscaper: you can adjust the hue of brown grass to make it green. The primary colors of the sky and lighthouse are lovely, but the brown grass distracts from that.

With a click and a drag, we can fix it. I'll use the direct adjustment tool, click in a brown portion of the grass, and drag the mouse until the grass is nice and green. As you can see, Lightroom adjusted the yellows, oranges, and greens to change colors while keeping them as natural as possible.

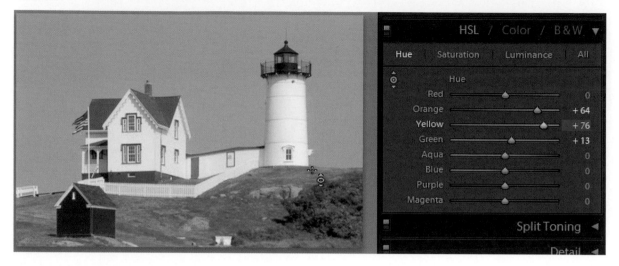

RESETTING CHANGES

If you overdo it, you can quickly undo all your changes by holding the **Alt** key (on a PC) or the **Opt** key (on a Mac) and then clicking **Reset Hue** (or **Reset Saturation** or **Reset Luminance**).

USING THE ALL AND COLOR TABS

The All and Color tabs do absolutely nothing different than the Hue, Saturation, and Luminance tabs; they simply provide a different user interface. For reference, here's the Color tab, which has you select a color and then adjust the hue, saturation, and luminance, rather than selecting the setting and choosing which color to change. The All tab lists all the Hue, Saturation, and Luminance adjustments, and is useful for reverse-engineering presets.

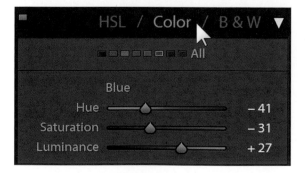

USING SPLIT TONING

Split Toning is a great way to add Instagram-like effects to your pictures. With Split Toning, you add a color tint to your picture. Optionally, you can add different tints to the highlights and shadows.

First, click the **Highlights** color box, as the following example shows with an unedited family snapshot.

Next, select a color tone to add to the highlights. The suggestions at the top of the Highlights panel are common choices. As you can see, my selection added a warm orange tint to the brightest parts of the picture.

Next, click the color box for the shadows. You can select the same color to tone the entire picture equally, or tone the shadows differently. In the next example, I'll add a cool tone to the shadows to complement the warm tone I added to the highlights.

These subtle changes give the picture an aged film look. Use the **Balance** slider to adjust how much of the picture Lightroom considers to be shadows or highlights. I dragged it to the right, so more of the picture is considered highlights, resulting in an overall warmer picture.

ADDING VIGNETTING

Vignetting is usually the darkening of the corners of your picture. Like black-and-white photography, it entered photography out of necessity, but remains an important artistic effect.

Historically, many lens designs (especially wide-angle lenses) don't distribute light evenly across the entire frame. Specifically, the center of the frame is brighter and the corners get darker—often two or three stops darker.

Wide-angle lenses that don't show vignetting are the most highly prized and high quality; however, vignetting adds a classic look to photos and draws your attention to the center of the frame, helping to reduce distractions in the same way as background blur.

THE DIFFERENCE BETWEEN NATURAL AND POST-CROP VIGNETTING

Modern lenses still exhibit vignetting, but your camera or the Lightroom raw processing have the ability to automatically eliminate any natural vignetting created by your lens design. That's generally a good thing; I'd always prefer my pictures as true to life as possible. We'll discuss the Camera Calibration panel later in this book.

In this chapter, we'll discuss using the Effects panel to add Post-Crop Vignetting. The problem with natural vignetting is that it darkens the corners of your original picture. If you then crop into the left corner, the vignetting won't be even in all four corners, making your picture lopsided. It's better to eliminate the natural vignetting and then use Post-Crop Vignetting to apply it as an effect.

VIGNETTING AMOUNT

To darken the corners, drag the **Amount** slider to the left. The next example shows the effect of -31 vignetting (on the right) as compared to the original photo (on the left).

Technically, the corners are darker. Visually, the effect draws your eyes to the center of the frame, narrowing your field of view. If you slide the **Amount** slider to the right, the corners become brighter. This never occurs naturally, so you should only use it as a special effect. The next example compares the dark vignetting from the previous example to light vignetting of +31.

VIGNETTING MIDPOINT

Let's explore the other sliders by way of example, using the duck photo as an example. In this first example, we see a **Vignetting** amount of -31 with a **Midpoint** of 50 (on the left) and 0 (on the right). As you can see, using a lower midpoint causes the vignetting to creep in closer to the center of the picture.

The next example shows the effect of using a higher **Midpoint** value, comparing 50 (on the left) to 95 (on the right). Higher midpoints move the vignetting out towards the corners.

VIGNETTING ROUNDNESS

By default, vignetting follows the aspect ratio of your picture. For example, these pictures are rectangular, so the vignette is oval shaped. You can use the **Roundness** slider to make it a perfect circle in the middle of the frame or make it almost as rectangular as the original picture. This next example shows the **Roundness** set to 100, making the vignette a perfect circle.

The next example sets the **Roundness** to -79 and the **Midpoint** to 14. The vignette becomes almost as rectangular as the picture itself, almost creating a frame.

VIGNETTING FEATHER

The **Feather** slider controls how soft the vignetting is. Keeping the same settings as the previous example, this screenshot shows the effect of sliding **Feather** all the way to zero. As you can see, the edges are now completely sharp.

VIGNETTING HIGHLIGHTS

The **Highlights** slider is set to 0 by default, and that's usually where you should leave it. Drag it to the right to have your vignetting impact highlights less than mid-tones and shadows. To understand the effect, apply it to a photo with highlights at the corners. The next example has highlights in the clouds in the upper left. In the left photo, the **Amount** slider is set to -62, darkening the corners. In the right photo, the **Amount** is the same, but the **Highlights** have been set to 100. As you can see, the only change is that the brightest parts of the picture are now even brighter.

VIGNETTING STYLE

There's one more vignetting adjustment: **Style**, at the top of the panel. By default, it's set to **Highlight Priority**, and that's the most natural looking. You can try the other two highlight styles, the effect of which will vary depending on your photo.

In this next example, we see the same picture with a vignetting **Amount** of -62. On the left, the **Style** is set to **Highlight Priority**. On the right, the style is set to **Paint Overlay**. Because of the exaggerated highlights in the corners of this photo, Paint Overlay actually looks more natural.

The next example compares Highlight Priority on the left to Color Priority on the right. Color Priority applies the vignetting effect slightly less to colorful areas of the picture, such as the upper-left corner on this photo of a never-ending Stockholm sunset.

ADDING GRAIN

Grain is the texture that you see in some film pictures. Grain occurs in film because film is actually made of small chunks of chemicals, and you can see those chunks when you look at a film picture that has been enlarged significantly or was taken with high-speed film.

Grain is an imperfection, a negative side effect caused by the limitations of the chemical processes involved in film photography. Digital photography eliminates that completely, and that's usually a good thing. But not always.

Like the scratches on an old record, grain conveys a sense of classic timelessness. It reminds us of that period from the late 1800s to the early 2000s when grain was the background to every photo, TV show, and movie. Grain adds texture, emotion, and a sense of history.

If a picture is too perfect, grain might be the answer. Let's start with a family snapshot of my daughter with her grandmother and great-grandmother. Before applying grain, I'll use Split Toning to add an aged effect. The original is shown on the left.

Now, I'll use the Effects panel to add Grain. The **Amount**, **Size**, and **Roughness** sliders each make the grain more visible when you slide them to the right. The next example shows a close-up of the previous photo with grain added.

You should increase the size of the grain relative to the size the picture will be viewed. If you apply small grain in a small picture, you'll never see the grain. If you plan for people to view the picture on their smartphone and you want the grain visible, you'll have to increase the size a great deal. If you plan to make a large print, smaller grain will be better.

There's no formula for applying grain; add and adjust to taste, and preview the picture in the final format.

SUMMARY

Much of Lightroom is strictly practical, but this chapter covered some of the most fun parts of the tool: color, split toning, vignetting, and grain. With these effects, you can support your picture's mood and emotion. Special effects can't make a great picture, but they can support a picture and make it even better.

One final bit of advice: have fun, but tread lightly. Go crazy with the effects, but before you share a picture, dial everything back 30%. With editing, less is more.

9
Black & White

To view the videos that accompany this chapter, scan the QR code or follow the link:

SDP.io/LR5Ch9

Black-and-white photography has a long history. It was born out of necessity; the first photographic chemicals reacted to any visible light hitting the film, and didn't react differently to red, blue, and green light. Thus, the film recorded the luminance of different parts of the image, with no color information.

Even after color film was developed, black-and-white photography remained popular for decades with professional photographers because the development and printing process was much simpler and less expensive. With black-and-white photography, photographers could complete the entire photographic process in their homes, including creating a print. Color photographs had to be sent to a professional lab, increasing cost and time, and reducing the control the photographer had over the final image.

In the digital era, there's no longer any practical need for black-and-white photography. In fact, because almost every modern camera records color images (one exception is the $8,000 Leica M Monochrom), it's now much harder to make a black-and-white image than a color image.

But there are many valid reasons to make black-and-white images:

- They eliminate the visual weight associated with bright colors, which can improve composition. A yellow sign in the background isn't as distracting in a black-and-white photo, and a child's polka-dotted shirt won't steal the attention from their eyes.

- They force the viewer to address the form, shape, and light of the image, without being distracted by the color. Often, I'll ask myself if the color is actually adding something to the image. If it's not, I'll convert it to black-and-white.

- They evoke a feeling of timelessness, connecting your image with actual film images and continuing the long tradition of landscape photographers like Ansel Adams and street photographers like Robert Frank.

- As prints, they might blend better into your décor. When displayed in a series in a home or gallery, black-and-white images tend to look more cohesive than color images.

Here's a quick way to find pictures that might look great in black-and-white. In the Library grid view, select potential candidates and then press V to convert them all to black-and-white. Now, de-select any images that look better in black-and-white by **Ctrl-clicking** them (on a PC) or **Cmd-clicking** them (on a Mac). Press V again to convert the remaining images back to color.

Creating natural-looking black-and-white photos is far more complex than simply pressing V to remove the colors; you must train yourself to think in black-and-white. That can take years of practice, but in this chapter, I'll begin teaching you with a variety of examples.

CREATING BLACK-AND-WHITE LANDSCAPES

In the Develop module, use the B&W panel to convert your image to black-and-white and make all adjustments. First, look at this image of Glacier National Park in color:

In this picture, the color adds very little to the picture, making it a good candidate for conversion to black-and-white. Landscape photography was established in black-and-white, in locations exactly like this, so desaturating can be considered traditional rather than cheesy.

Clicking B&W desaturates the image and applies Lightroom's default Black & White Mix, as shown next. As you can see, Lightroom doesn't consider all colors equal; the warm colors are less bright, while the cool colors are more bright. This, however, is just a starting point.

Tip: Color-correct your photo before converting it. Having accurate colors will make it easier to adjust the black-and-white mix.

Now, I can begin the process of tweaking the brightness of different colors. The direct adjustment tool in the upper-left corner of the B&W panel is the most accurate way to modify specific parts of your picture.

I wanted to see more contrast in the sky, so I clicked the direct adjustment tool and then clicked the darker part of the sky (which had been blue). While holding down the mouse button, I dragged my cursor down to make it darker. As shown next, Lightroom detected that I clicked on blues and aquas and adjusted those colors in the picture.

Unfortunately, the adjustment also darkened the focal point of the image: the lake. I needed the lake to have contrast with the bare earth of the mountain top in the foreground. If I had tried to use the B&W panel, it would have changed all blues in the picture, including the sky.

The Adjustment Brush is the right choice for making a localized change like this. With the **Show Selected Mask Overlay** checkbox selected, I highlighted the lake, as shown next.

Next, I adjusted the **Adjustment Brush Exposure** slider until the lake had a unique mid-tone, between the darker mountains and the bright snow.

Before finishing any picture, especially black-and-white photos, you should look at the histogram. Ideally, the histogram should touch both the left and right sides. This is important for color photos, but it's critical for black-and-white photos, because contrast is critical. Closing the Adjustment Brush, I increased the **Whites** and decreased the **Blacks** until the histogram touched both edges. With these adjustments, the final image is distinctly better than the original.

TURNING OFF THE AUTO-MIX

By default, Lightroom automatically drops the warm colors and raises the cool colors when converting to black-and-white. I like this, but if you don't, you can turn it off.

Open the **Edit** menu (on a PC) or the **Lightroom** menu (on a Mac) and select **Preferences**. Then, select the **Presets** tab and clear the **Apply Auto Mix** checkbox, as shown next. Lightroom won't change any of the photos you've already imported, but new pictures that you convert to black-and-white will use a flat mix, with all colors set to 0.

Preferences

General | Presets | External Editing | File Handling | Interface | Lightroom mobile

Default Develop Settings

☐ Apply auto tone adjustments
☑ Apply auto mix when first converting to black and white
☐ Make defaults specific to camera serial number
☐ Make defaults specific to camera ISO setting

CREATING BLACK-AND-WHITE WEDDING PHOTOGRAPHS

Wedding photographs are often converted to black-and-white to add timelessness to the image. It's rarely prudent to shoot film for modern wedding photography, however; it's too easy to miss shots like this next example while changing film.

After clicking **B&W**, the default mix isn't bad. It's a matter of preference, but the low-contrast look is popular.

More contrast would look better in a print, however. Because the color image is almost monochromatic (it was lit entirely by the setting sun), moving the individual sliders isn't effective. I'll need to rely on the adjustments in the Basic panel, instead.

The Contrast slider isn't always the best way to add contrast. For this photo, I need to move the entire histogram left to fill in the darks and shadows, and the easiest way to do that is to slide the Blacks and Shadows sliders left in the Basic panel, as shown next.

As you can see, the histogram is spread out through almost the entire range.

CREATING BLACK-AND-WHITE STREET PHOTOGRAPHY

Color is a powerful element in this street photography example. In fact, the color is overwhelming, distracting from the subject of the photo, which is the interaction between the walking couple and the two-dimensional musicians.

Clicking B&W, the default black-and-white mix is surprisingly dark and low contrast.

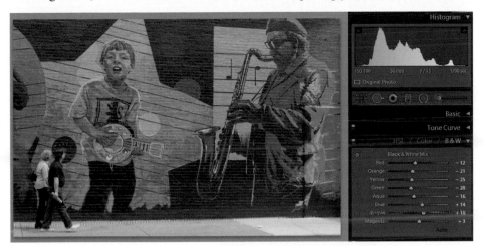

Starting with the Basic panel, I adjusted the **Exposure**, **Contrast**, **Shadows**, **Whites**, and **Blacks** to increase contrast in the picture.

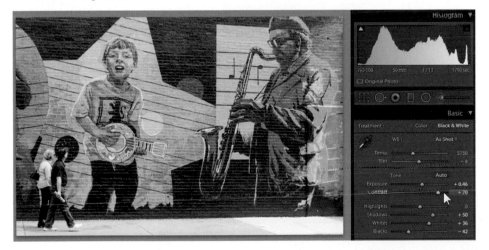

Now that the photo has sufficient "pop," I can begin to examine the visual weight of the photo to make sure that the subject has sufficient contrast. Unfortunately, the man in the foreground is blending into the dark background, so I need to separate him further. I use the direct adjustment tool in the B&W panel to adjust the brightness of his shirt and pants. That also adjusted some of the colors on the wall, but I liked the look of it.

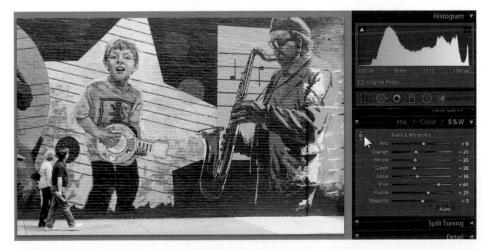

CREATING BLACK-AND-WHITE PORTRAITS

In this candid portrait, my daughter has a nice expression and she's in good light. However, the colors on her shirt are distracting. In fact, even the bright greens of the trees in the background are distracting. With portraiture, it's very easy for colors to overwhelm the subject of the photo, which is usually the model's expression.

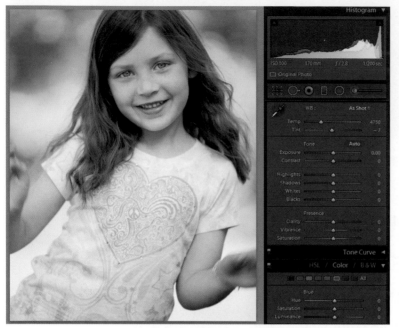

Converting it to black-and-white definitely helps, but the default mix (shown next) left her face a bit unnatural. Skin tones almost always need adjustment when converting to black-and-white. In this example, her skin tones are very grey, but they would look better closer to white.

To make the skin tones brighter, I again used the direct adjustment tool in the B&W panel, and grabbed several different spots on her face.

That's an improvement, but her shirt is still brighter than her face. Her shirt literally has every color in it, so I won't be able to use the B&W panel. Instead, I'll once again resort to the Adjustment Brush. As shown next, I selected the shirt (using **Auto Mask** to select the edges, and then disabling the option to select the inside of the shirt) and then reduced the exposure and contrast.

Her eyes are still a bit dark, so I created a new adjustment brush and painted over her eyes. Then, I raised the exposure a small amount.

In the final product, the skin tones are natural, her eyes are bright, and nothing in the photo has more visual weight than her expression.

SUMMARY

Today, black-and-white photos are usually created with post-processing effects, and that's ideal. It might seem counter-intuitive, but it's far better to shoot a black-and-white photo in color, because you can use the color information when making adjustments in Lightroom.

Black-and-white processing has many unique concerns. Without color information, the contrast in the image becomes much more important. You should carefully examine the histogram and ensure that the photo has both pure white and pure black in it, especially if you plan to make a print. Localized contrast is also very important; contrasty parts of your picture can steal attention away from your subject.

You can't rely on different colors to separate a subject from background elements, either. In the landscape example, I had to change the brightness of the lake to prevent it from visually blending into the earth. In the street photography example, I had to adjust the brightness of the subject's clothing to separate them from a differently-colored background.

10
Sharpening & Noise Reduction

To view the videos that accompany this chapter, scan the QR code or follow the link:

SDP.io/LR5Ch10

This chapter covers two techniques for improving image quality: sharpening and noise reduction. Sharpening exaggerates the appearance of details in your pictures, making them seem sharper. Noise reduction reduces the speckles and grain found in all digital photos, but makes pictures less sharp.

While the default settings work for many pictures, some pictures benefit from tweaking. This chapter will show you the effects of each slider and give you recommendations about how and how not to use each setting.

SHARPENING

The term "sharpening" is incredibly misleading. Of course, you can't actually extract more detail from an image than your camera and lens originally captured. Unless, of course, you're in an action movie and yell, "ENHANCE!"

In other words, sharpening won't make a blurry picture sharp. If you missed focus, you'll just need to re-take the picture. If you had camera shake in an image, Photoshop has a tool that might help, but the picture is probably a loss.

Sharpening does give the illusion of showing a sharper picture by accentuating areas of contrast in an image. All cameras automatically add some artificial sharpening to a picture when taking JPG images, and Lightroom adds some sharpening to an image when you import a raw file. For many pictures, you might want to add more sharpening.

Sharpening is more important to wildlife photography than any other type of photography because people value the details and texture in animal photos. Take this close-up of a house finch; it's perfectly in focus with no camera shake, and it's reasonably sharp with Lightroom's default sharpening of 25.

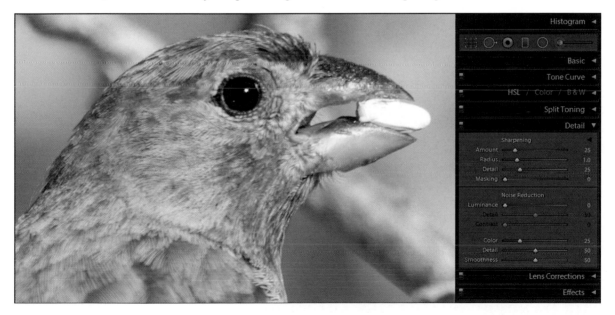

If I open the Detail pane and drag the **Sharpening** all the way to the right, to 150, it definitely looks sharper. Your eyes now easily see the outline of every feather. Yet, there's not actually any more detail; Lightroom has simply increased the contrast between each feather.

Zooming in to 4:1 (so that each individual pixel is 4 times bigger than full size) we can more easily see what Lightroom did. In the before picture, with Sharpening at 25, each pixel smoothly blends into the next. With the Sharpening at 150, the difference between highlights and shadows is much more distinct. However, every line in every feather in the sharpened picture is visible in the original picture. Sharpening added no detail; it just made the details more obvious.

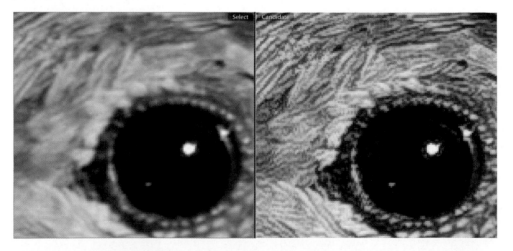

THE DANGERS OF OVER-SHARPENING

Over-sharpened pictures look unnatural, and most people can sense that even when they don't know what sharpening does. For example, most TVs are over-sharpened by default; TV manufacturers do this to make the TVs look better in the showroom than other TVs. Yet, when you get it home, the over-sharpening actually makes the picture quality worse. Lowering your TV's sharpening will give a more natural look.

Choosing the correct amount of sharpening is more art than science; simply adjust the slider until it looks the best to you, and then reduce it 30%. Often, when you're gradually adjusting the sharpening up, it seems like the photo looks better with more and more sharpening, but it's more obvious to the viewer who hasn't been participating in the editing process.

SHARPENING RADIUS

The Radius setting tells Lightroom how many pixels to look around each pixel for contrast. Generally, the default of 1 is best for digital files; using a higher radius is more useful with film.

Using a radius lower than 1 can actually add artificial detail to the image, seemingly doubling details. Using a radius higher than one eliminates some detail, grouping together areas of contrast. The next example shows maximum sharpening with a radius of 1 (on the left) and a radius of 3 (on the right). As you can see, the photo on the right lost much of the detail in the feathers, because Lightroom looked three pixels around each pixel for areas of contrast, causing fine details to be grouped together.

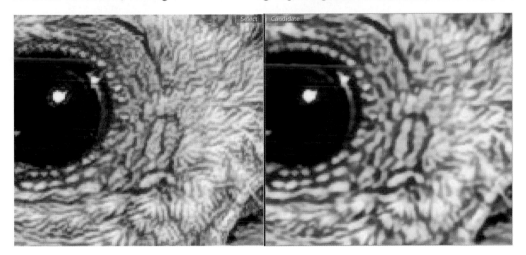

SHARPENING DETAIL

In practice, the Detail slider acts like an extension to the **Sharpening Amount** slider. Move it to the right to add more sharpening.

In the following side-by-side example, you see the same picture with different settings. One of them is 100% **Sharpening Amount** and 50% **Sharpening Detail**, and the other is 100% **Sharpening Detail** and 50% **Sharpening Amount**. The end result is identical.

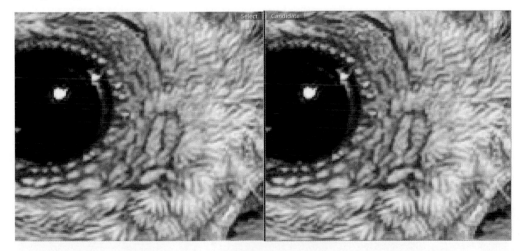

Therefore, you only need to increase the Detail slider if you want to add more sharpening than is possible using the **Sharpening Amount** slider.

SHARPENING MASKING

You often want to add sharpening to the subject of your photo, but you never want to add sharpening to the out-of-focus areas of your photo. When you do, you'll accentuate noise.

To illustrate this, let's take a close-up look at the background of the same photo. The photo on the right has Lightroom's default sharpening. On the photo on the right, I've raised the **Sharpening Amount** and **Detail** sliders all the way.

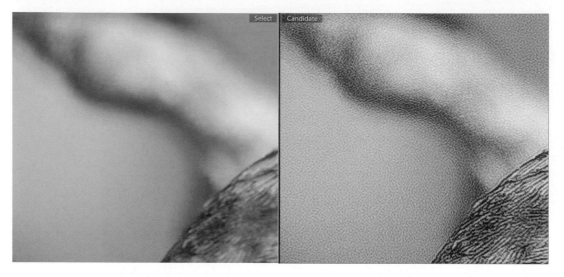

As you can see, over-sharpening makes the creamy background look grainy and unnatural. Fortunately, Lightroom's Masking feature allows you to apply sharpening just to the detailed parts of your picture, without impacting the blurry background.

To use the **Masking** slider, hold down the **Alt** or **Opt** key on your keyboard while adjusting the slider with your mouse. Lightroom will display a black-and-white mask of your image. The white parts of the image are the only parts Lightroom will apply sharpening to. As you slide **Masking** right, more of your picture will become black, revealing just the most detailed parts of your picture. Slide **Masking** until the blurry background is all black, but the detailed parts of the image are shown in white.

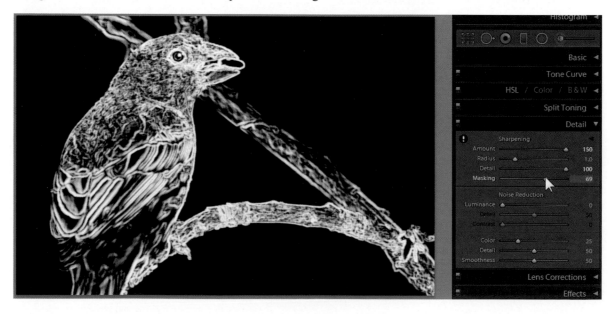

The next side-by-side example shows Lightroom's default settings compared with a massively over-sharpened image using masking. As you can see, the over-sharpened image on the right still has a creamy background, with no visible noise.

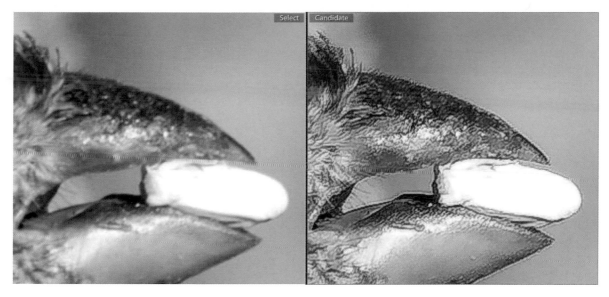

NOISE REDUCTION

Light is a bit more random and sparse than it seems to your eye, and as a result, cameras don't capture your world perfectly.

Noise is to pictures what static is to old records and tapes; it's a bit of artificial impurity. Every camera and every picture has some noise. The less light your camera gathers for a picture, the more noise it has.

Higher ISOs create the same brightness of picture with less total light. Therefore, the images are noisier than pictures taken at lower ISOs. Cameras with smaller sensors create pictures just as bright as cameras with bigger sensors, but they use less total light to make those pictures. Therefore, images taken with smaller sensors have more noise than images taken with a bigger sensor when the settings are the same.

When you take a JPG picture, your camera automatically applies some noise reduction. When you shoot raw, Lightroom applies the noise reduction for you. Some noise reduction almost always improves the appearance of a picture.

Noise reduction has an unwanted side-effect, however: it reduces detail and sharpness. Therefore, you must carefully manage the amount of noise reduction in a picture to reduce ugly noise while minimizing the loss of sharpness.

A NOISY EXAMPLE

This photo of a Pygmy Slow Loris was taken at ISO 25,600 in a dark cave habitat. Because of the high ISO, it's very noisy. Notice the cursor is pointing to the **Detail Zoom** tool, which allows you to click a portion of the picture to view it at 100%. Zoomed out, the picture doesn't look that noisy, but in the Detail view, you can see that none of the textures are smooth. Instead, the noise seems to make the fur speckled.

For the sake of this example, I'll raise the **Noise Reduction Luminance** slider to 100. As the next example shows, the noise is completely gone. However, the entire picture now looks very soft and waxy. That's what a picture looks like when the noise reduction is too high.

The art of using noise reduction is finding the perfect balance between 0 (with the most noise and detail) and 100 (with the least noise and detail). For most pictures, you only need to adjust the **Luminance** slider; you can leave the remaining sliders at their default settings.

I wish Noise Reduction had the same Masking tool that Sharpening has; it would be nice to reduce the noise in the background without reducing the detail in the subject. Unfortunately, you can't mask noise reduction in Lightroom. However, if you bring the photo into Photoshop, you can separate the foreground and background into different layers, and apply more noise reduction to the background.

LUMINANCE VS. COLOR NOISE REDUCTION

Lightroom has two separate sets of noise reduction sliders: **Luminance** and **Color**. This example will illustrate the difference. The left example has the Color noise reduction raised all the way up, thus showing you only luminance noise. The right example has the Luminance noise reduction raised, thus showing you only color noise.

If your image appears speckled (as in the left image in the previous example), that can be fixed by raising the **Luminance** slider. If your image has randomly colored spots in it when zoomed in (as in the right image), you'll want to rise the **Color** slider. Usually, the default Color setting is fine, however.

THE DETAIL SLIDERS

Both Luminance and Color noise reduction have detail sliders, which try to recover some of the detail lost in the noise reduction process. You don't usually need to adjust this slider, but you might be able to improve the appearance of a picture with heavy noise reduction by raising the **Detail** slider. The next two examples show the same photo with the **Luminance Detail** slider at 0 and 100 to exaggerate the effect. The default is 50.

The difference is subtle, but the image on the right has more contrast. Unfortunately, it also shows more noise when you look close. Raising **Detail** undoes some of the effect of reducing noise.

You shouldn't need to adjust the **Color Detail** slider, but if you do, Lightroom will exaggerate differences in color between nearby pixels.

THE CONTRAST SLIDER

Contrast has a similar effect to Detail; it adds a bit more pop to offset heavy noise reduction. If you raise the **Detail** slider all the way and want even more pop, raise the **Contrast** slider. As with the **Detail** slider, raising the **Contrast** slider undoes some of the effect of noise reduction.

The following example shows the same photo with the **Contrast** slider at 0 (on the left) and at 100 (on the right).

THE SMOOTHNESS SLIDER

By default, Lightroom sets the **Smoothness** slider to 50, and that's fine for most pictures. Sliding it to 0 will reduce Lightroom's color noise reduction effect, showing random spots of colored pixels in the photo. Sliding it to 100 will maximize the color noise reduction effect, reducing areas of detail in the photo.

The next example shows the same photo with the Smoothness at 0 (on the left) and at 100 (on the right).

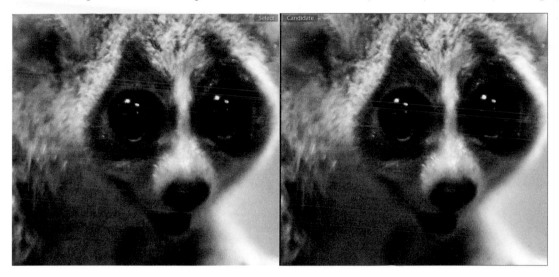

On the left, the random reds and greens are very distracting and unnatural. On the right, the high color smoothness has made parts of the picture completely black-and-white. Neither extreme is ideal; something in between is almost always better, and the default of 50 works fine for most images.

SUMMARY

For most pictures, Lightroom's default sharpening and noise reduction are fine. If you plan to make a large print, or you need to crop an image heavily (thus showing every pixel) you can adjust the sharpening and noise reduction to improve the visible image quality.

Be careful about overdoing either setting. Pictures might look better the higher you raise the sliders, but they'll look less natural, too. Most people are accustomed to seeing somewhat unsharp, noisy pictures, and won't think anything of it. Think of sharpening and noise reduction like plastic surgery; most people never mention flaws in someone's face or body, but if someone gets too much plastic surgery, it becomes all people can think about.

11
Lens Corrections & Camera Calibration

To view the videos that accompany this chapter, scan the QR code or follow the link:

SDP.io/LR5Ch11

Lenses and cameras are never perfect. Colors might be slightly off, subjects might be distorted, and fringing can give purple and blue edges to high-contrast subjects.

Remarkably, Lightroom can automatically fix many of these problems. The beginning of this chapter will show you how to automatically apply those corrections to all new pictures. If you're in a hurry, you can just read the next two sections and skip the rest of this chapter.

If you want to know how to straighten the lines of buildings, or you're a perfectionist about distortion and chromatic aberration (that weird fringing you see at the edges of backlit pictures), read the entire chapter and practice with your own images to master these tools.

LENS CORRECTIONS

The lens corrections panel is one of the most important in Lightroom because it can quickly and significantly improve the quality of your photos, making inexpensive lenses give the results of professional gear. It's also one of the easiest to use: simply select all three checkboxes on the Basic tab, and apply that to all of your pictures.

Most photographers will never need to explore the other tabs; Lightroom's automatic processing works wonderfully. However, for the curious, I will describe how to use every panel. First, let me show you how to apply automatic corrections to every picture that you import.

APPLYING LENS CORRECTIONS AUTOMATICALLY

Here's something everyone should do: configure Lightroom to automatically apply lens corrections to newly imported pictures.

First, select any picture in your catalog, select the Develop module, and open the Lens Corrections panel. Select all three checkboxes, as shown in the previous screenshot. Now, make a new preset by clicking the + symbol on the Presets panel, as shown next.

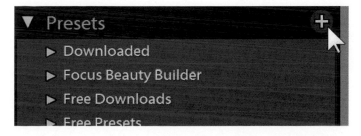

In the New Develop Preset dialog, set the **Preset Name** to **Lens Corrections**. Click the **Check None** button at the bottom of the dialog to clear all checkboxes, and then select the **Lens Corrections** checkbox, as shown next. Click **Create**.

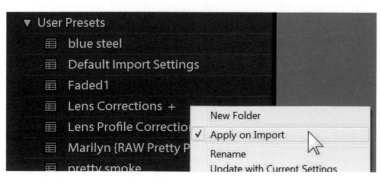

I cover presets in more detail elsewhere in the book, but a preset applies a group of settings to pictures, and you can apply a preset to pictures automatically when you import them.

Now that you've made the preset, you just need to configure the Import dialog to apply it automatically to new pictures. Right-click the new Preset (you can find it under User Presets) and then click **Apply On Import**, as shown next. Lightroom will show a + symbol next to the preset name to indicate that it's the default preset.

The next time you import pictures, you'll see your preset selected under Apply During Import.

THE BASIC TAB OF THE LENS CORRECTIONS PANEL

The Basic tab has three checkboxes and a handful of buttons. First, I'll describe the three checkboxes, and then the buttons.

ENABLE PROFILE CORRECTIONS

This enables Lightroom to fix the natural distortion and vignetting of your lens, as set on the Profile tab. Basically, Lightroom will try to detect your lens from the metadata and then remove darkened corners and warping. All lenses have some degree of these problems.

The following example shows the most extreme improvement I could find, and even with this before-and-after, you'll have to look closely to notice the differences. With profile corrections, the picture on the right has less vignetting, so the corners are a bit brighter (and more natural). Lightroom removed the distortion, too, so the straight lines in the tower are now exactly straight, instead of very slightly curved.

The best way to see the examples for yourself is simply to select a picture and turn profile corrections on and off repeatedly. You'll see a more significant difference at wider angles and with less expensive lenses.

REMOVE CHROMATIC ABERRATION

Chromatic aberration is an odd coloring that you get along the edges of high-contrast subjects. For example, when zoomed in to 100%, one side of a tree branch might have a purple edge, while the other side has a green edge.

Almost all lenses show some chromatic aberration. It's caused by the fact that different wavelengths of light travel through the lens optics differently. For example, in the following close-up of the same Eiffel tower picture, the tower itself is a neutral color, which means it was reflecting red, green, and blue frequencies towards my lens. Just above the hand in the before picture on the left, you can see that the green frequencies hit the sensor slightly to the left, and the red frequencies hit the sensor slightly to the right.

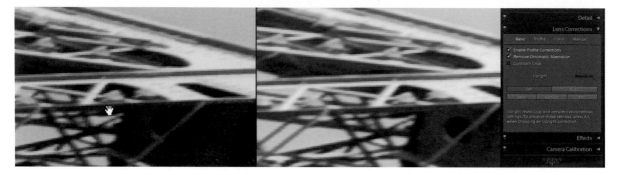

You don't have to understand the physics behind chromatic aberrations to fix them. Simply select the **Remove Chromatic Aberration** checkbox on the Basic tab of the Lens Corrections panel, and Lightroom does a great job of fixing it, as shown by the right image on the previous screenshot. I recommend always selecting this checkbox because I've never seen a negative side effect of it.

Notice that there are some other differences between the before and after images in the previous screenshot. For example, the circle (to the right and slightly below the hand cursor) has been moved closer to the lower-right corner, and the shadows are brighter. These adjustments were caused because the **Enable Profile Corrections** checkbox is also selected.

CONSTRAIN CROP

There's no reason not to select this checkbox, though it rarely factors into your final image. If you do make severe adjustments, as demonstrated by the two examples below, selecting **Constrain Crop** (shown in the second example) prevents the background from appearing in your final photo by creating a tighter crop that shows only your original image.

THE LEVEL, ASPECT RATIO, AND PERSPECTIVE CORRECTION BUTTONS

At the bottom of the Lens Corrections panel, Lightroom provides five buttons that can automatically straighten your photos and even correct some distortions.

Unlike the **Profile Corrections** and **Chromatic Aberration** checkboxes, these buttons aren't especially reliable. They probably work well half the time, but it's so easy to level pictures yourself that I generally recommend doing it manually with cropping. To correct distortions, use the Manual tab instead (described later in this chapter).

The five buttons are:

- **Level.** Clicking this button tries to rotate your picture to make it level. Sometimes it's right, but it's often very wrong. You can also use the crop tool and rotate the picture manually.

- **Vertical.** This button levels your picture and attempts to correct for vertical distortions. Vertical distortions are caused when you're pointing a lens upwards, for example, when you're taking a picture of a building with a wide-angle lens from the ground.

- **Full.** Full enables both **Level** and **Vertical**, and adds horizontal correction. You might want horizontal correction if you took a picture of a big subject (like a building) at an angle. In that scenario, the far corner of the building would be smaller than the near corner. Full can correct this for you.

- **Auto**. Like pulling the arm on a slot machine, you're probably going to lose when you click **Auto**. Sometimes, though, Auto does improve your picture. It's usually more intelligent than **Full**.

- **Off.** This button turns these settings off.

This example shows the **Auto** button adjusting the vertical lines in a redwood forest:

This example shows the **Vertical** button adjusting the same picture. As you can see, the button attempts to correct for the distortion caused by looking upwards at these subjects. Instead of the faraway parts of the trees being smaller, they're now equally as wide at the top as the bottom. It's as if I took the picture level with the trees, rather than pointing up at them:

Here are examples of the tools going rogue and destroying an image. First, the **Level** button takes the almost-level original image and twists it in a very strange way:

Next, the **Full** button really does eliminate the distortion in Notre Dame, straightening the towers to perfect vertical. But, rather than looking more correct, the image looks completely unnatural, as if it were leaning towards the viewer.

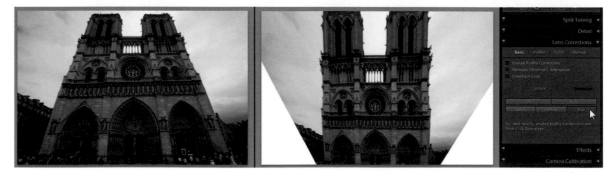

There are appropriate times to use these buttons, but generally, you're better off using the Manual tab.

THE PROFILE TAB OF THE LENS CORRECTIONS PANEL

You might never have to look at the Profile tab, shown next. Selecting the **Enable Profile Corrections** checkbox on the Basic tab selects the same checkbox on the Profile tab without you ever opening the tab. Typically, you can simply select **Auto**, and Lightroom will examine the photo's metadata to discover which lens you used and use a profile designed specifically for that lens.

You might have to use the Profile tab if your camera didn't record your lens metadata. This might happen if you're using an adapter to connect your lens to your camera. If that's the case, simply select **Custom** from the **Setup** list and manually select your lens Make, Model, and Profile.

If you can't find your lens, you can either create your own custom profile or download a profile that someone else has made. Visit *sdp.io/CustomProfile* and use the Windows or Mac tools to download or create your own profiles. The next figure shows the Adobe Lens Profile Downloader, which has a couple hundred common profiles.

Or, for a quick fix when Lightroom can't adjust your lens automatically, simply slide the **Distortion** and **Vignetting** sliders until your image looks natural. People are accustomed to seeing both distortion and vignetting in photos, so it's usually not a problem if you don't correct them perfectly.

THE COLOR TAB OF THE LENS CORRECTIONS PANEL

Simply selecting **Remove Chromatic Aberration** on the Basic tab is good enough to solve most fringing problems. However, if you still find annoying chromatic aberrations, you can fix them using the Color tab of the Lens Corrections panel.

First, select the color of the fringing that you want to remove using the dropper tool, as shown next. You'll usually need to select two different colors, which will be visible on the left and right sides of high-contrast subjects.

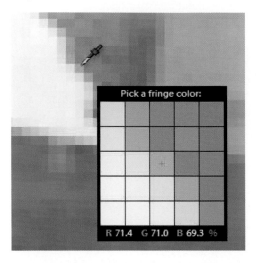

You can also manually adjust the sliders to change the color ranges that Lightroom looks for when correcting aberrations. For example, while fringing is usually either purple or green, some lenses show purple fringing closer to blue, while others are closer to magenta. Use the **Purple Hue** and **Green Hue** sliders to select the color ranges of your fringing. Slide the **Amount** sliders to the right only as far as you need to in order to remove all annoying aberration. If you slide it too far, you can actually add strange colors to the rest of your image

The next example shows green fringing removed from a close-up of the previous backlit image of the arch.

THE MANUAL TAB OF THE LENS CORRECTIONS PANEL

The Manual tab allows you to correct perspective distortion. Perspective distortion occurs because subjects that are faraway look smaller than close-up subjects, both to your camera and to your eyes. The more wide-angle your lens, the greater the perspective distortion.

Perspective distortion rarely requires correction. Our brains expect the top of a building to be smaller than the base when a picture is taken from ground level because that's how we see buildings in person. Sometimes, such as when creating architectural photography, you might want to correct this distortion.

Here's a brief overview of the different sliders, followed by some examples:

- **Distortion**. Corrects the pincushion effect caused by the design of your lens. Usually it's more accurate to use the automatic correction provided by the Profile tab.

- **Vertical**. Corrects perspective correction caused by tilting the camera up at a big subject (or, less frequently, down at a subject).

- **Horizontal**. Corrects perspective distortion caused by tilting the camera sideways to a big subject, such as when you're taking a picture of a square building but you're standing slightly off-center.

- **Rotate**. Rotates your image to make it level.

- **Scale**. Zooms the crop on your photo.

- **Aspect**. Squeezes your photo horizontally or vertically.

- **Lens Vignetting**. Corrects darkened corners.

This first example shows the **Horizontal** slider being used to correct for the fact that I was standing slightly off center of this arch outside the Roman Forum. As you can see, the left side of the original picture is a bit more prominent than the right side, showing that I was standing just left of center. The **Horizontal** correction fixed it perfectly:

The next example revisits the Notre Dame picture that Lightroom's automatic settings mangled. By manually adjusting the settings and cropping the picture, I was able to create a natural-looking image of Notre Dame's architecture.

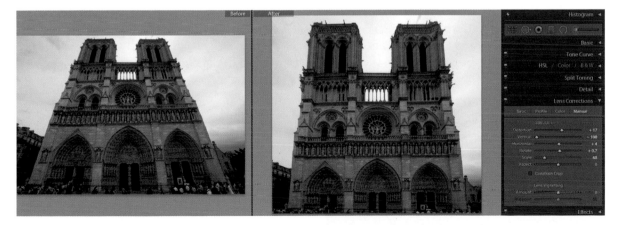

CAMERA CALIBRATION

The Camera Calibration panel, shown next, allows you to make basic adjustments to color that are specific to each camera model you might use. You shouldn't ever need to touch it; however, if you find that you always need to adjust the colors for a specific camera, this is the right place to make those adjustments.

The Process list switches between different versions of the Lightroom's raw processing software. The original raw processing software was created in 2003. Adobe improved it in 2010, and then again in 2012.

If you've been a Lightroom user for many years, you might have pictures that were processed in earlier versions of the software. If Lightroom automatically updated the process for those photos, your pictures would look slightly different, and Lightroom never changes your pictures without your permission. Therefore, to update the process to the latest version, click the list and select **2012 (Current)**. You won't see a drastic change in your photos. The differences are subtle, but I generally prefer the current process.

You can also update the process from the Library module, as shown next. If you want to update all your pictures, select the Library module, press **Ctrl+A** (on a PC) or **Cmd+A** (on a Mac), right-click click an image, click **Develop Settings**, and then select **Update To Current Process**.

SUMMARY

Lens corrections and camera calibrations are best used with Lightroom's automatic settings. However, when you find an image that's worthy of editing to perfection, it's important that you know how to manually adjust the settings to fix odd colors, distortions, and perspective problems.

12

Presets, History, Snapshots, & Before & After

To view the videos that accompany this chapter, scan the QR code or follow the link:

SDP.io/LR5Ch12

Every change you make to a picture can be undone. Press **Ctrl+Z** (on a PC) or **Cmd+Z** (on a Mac) to undo your last change, or press it repeatedly to undo multiple changes. If you go too far and need to redo a change, press **Ctrl+Shift+Z** (on a PC) or **Cmd+Shift+Z** (on a Mac).

This ability to move forward and backwards through your edits is much more powerful than simply undoing changes. The History panel lets you view your edits and jump to any change with a single click. You can use Snapshots to save your current state, and return to it at any point later if you make changes you don't like. Finally, the Before & After tool compares your original picture to your edits, so you can see the improvements you've made (and make sure you haven't overdone them).

BEFORE & AFTER

The Before & After tool is a great way to see exactly how much you've improved your picture. Sometimes, you'll see that you actually made part of your picture worse, and you might decide to go back and undo some changes.

If the toolbar is visible, click the **Before & After** tool (shown below the cursor in the next example) or press **Y** on your keyboard. Press **Y** again to return to the original view. If the toolbar isn't visible, press **T** to view it.

Lightroom provides several different viewing modes, which you can select by clicking the triangle to the right of the **Before & After** button, as shown next. The **Left/Right Split** is usually the most efficient use of your screen space, but the **Top/Bottom Split** is useful for panoramas.

To switch between a side-by-side and split view, press **Shift+Y**. To switch from Left/Right to Top/Bottom, press **Alt+Y** or **Opt+Y**. The Split modes show half the Before image and half the After image, as shown next.

There are three other buttons on the Before & After toolbar. From left to right, they are:

- **Copy Before's Settings To After**. If you edit your picture but the Before image still looks better, click this button to undo all your changes. If the Before picture shows the unedited photo, this resets all your changes to the original settings when you imported the picture.

- **Copy After's Settings To Before**. Click this to copy your settings from the After picture to the before. Now, you can make additional changes and compare those additional changes to the image's status at the point you clicked this button, rather than to the original, unedited image.

- **Swap Before And After Settings**. This copies the settings from Before to After, and from After to Before. Essentially, it undoes all of your changes and allows you to compare any future changes to your current edit.

The first time you use the Before & After tool on a picture, the Before will be the original, unedited picture. However, if you use the above buttons, the Before might show an edited version of the picture. This is useful for viewing your progress compared to a partially edited picture, rather than to the original unedited picture.

These buttons do not behave like the Compare tool in the Library module; the Before will always be on the left, and the After will always be on the right. Instead, they actually change the settings applied to the different versions of the picture.

You can also use the History panel to change the point in your edits shown in the Before picture. Simply click a step in the History and then select **Copy History Step Settings To Before**, as shown next.

HISTORY

The History panel on the left of the Develop module shows a list of every edit you've made to your picture. There's not much too it; click any edit to return to that state.

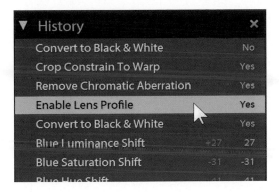

SNAPSHOTS

I often make dozens of changes to a picture and then realize that the last ten minutes of my changes actually made my picture worse. If you plan to experiment with a picture, it's a good idea to make a snapshot of the picture's current state.

To create a snapshot, click the + symbol on the Snapshot panel. Then, type a name for your snapshot.

You can also create a snapshot by right-clicking a step in the History panel, as shown next:

Once you've made a snapshot or two, you can click a snapshot to return to those settings. You can also right-click a snapshot to copy it to the Before image (when using Before & After), rename it, delete it, or replace the snapshot with your current edits.

PRESETS

Snapshots are only visible within a single picture; you can't switch pictures and then select a snapshot. If you want to apply settings to different pictures, either synchronize them, copy and paste them, or create a Preset.

Presets are simply a collection of settings, but they're incredibly useful for creative editing. For example, I have presets for different black-and-white mixes, sepia toning, and creative (Instagram-like) effects.

I don't necessarily use the presets exactly as they are, but I often browse through my presets for creative inspiration. During the process of browsing, I often see an effect that catches my eye. I might start with that preset and edit it further to make the picture unique. Or, I might combine the best effects of two different presets.

Lightroom has some presets built in. You can also download presets from the Internet or create your own.

CREATING YOUR OWN PRESETS

To create your own preset, edit a picture and click the Plus symbol on the Presets panel, as shown next.

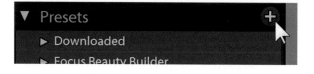

Lightroom opens the New Develop Preset dialog. Here, you can name your preset, store it in a specific folder, and choose which settings you want to make part of the preset. Click **Create** when you're ready.

New Develop Preset

Preset Name: Rich Colors

Folder: User Presets

Auto Settings

☐ Auto Tone

Settings

☑ White Balance	☑ Treatment (Color)	☐ Lens Corrections
		☐ Lens Profile Corrections
☑ Basic Tone	☑ Color	☐ Chromatic Aberration
☑ Exposure	☑ Saturation	☐ Upright Mode
☑ Contrast	☑ Vibrance	☐ Upright Transforms
☑ Highlights	☑ Color Adjustments	☐ Transform
☑ Shadows		☐ Lens Vignetting
☑ White Clipping	☑ Split Toning	
☑ Black Clipping		☑ Effects
	☑ Graduated Filters	☑ Post-Crop Vignetting
☑ Tone Curve	☑ Radial Filters	☑ Grain
☑ Clarity	☑ Noise Reduction	
	☑ Luminance	☑ Process Version
☑ Sharpening	☑ Color	☑ Calibration

Check All   Check None   Create   Cancel

By default, Lightroom stores your custom presets in the User Presets folder. If you make more than a couple presets, you'll want to make a custom folder. Simply select **New Folder** from the **Folder** list, as shown next:

You can update a preset at any time by right-clicking it and then selecting **Update With Current Settings**.

APPLYING PRESETS

Once you create a preset, simply click it to instantly apply those settings to the current photo.

Lightroom includes several dozen presets built in. My favorite way to use these is to open the Navigator panel (the top-most panel on the left side of the Develop module) and then hover my cursor over different presets. The Navigator panel will preview the preset, giving me a quick idea of the preset's effect without committing any changes. The next example shows the Navigator panel previewing the **Lightroom B&W Toned Presets\ Sepia Tone** preset.

INSTALLING DOWNLOADED PRESETS

You can download presets from the Internet. It's very brute force to use settings someone else created for a different picture, but nonetheless, presets are fun to use for creative inspiration or a starting point for your own editing.

Simply search the Internet for "free Lightroom presets," and you'll find dozens of websites with presets available. Once you download them, follow these steps to install them:

1. Download the presets to your computer. If the file has a .zip extension, open it and copy the files to a folder.

2. In Lightroom, right-click the **Presets** folder in which you want to store the new presets inside, and click **Import**, as shown next. You can also use the right-click menu to create a new folder.

3. Browse to the folder you used to store your presets, select them (or **Shift-click** to select multiple presets), and click **Open**.

That's it! Click any preset to apply it.

Many websites will attempt to sell you presets. Don't bother; there's nothing magical about these paid presets; they're simply a collection of settings. There are plenty of freely available presets with similar settings.

EXPORTING PRESETS

You can export presets to give them to other people or to copy them between different catalogs. To export a preset, simply right-click it and then click **Export**. Lightroom will create a .lrtemplate file that anyone can import using the instructions in the previous section.

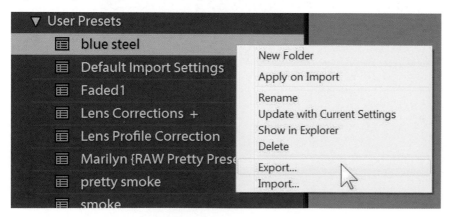

Exporting a preset creates a new copy of it. You can also find the original copy of your preset by right-clicking it and then selecting **Show In Explorer** (on a PC) or **Show In Finder** (on a Mac).

STORING PRESETS WITH A CATALOG

By default, Lightroom uses a single set of presets for all catalogs. If you create multiple catalogs and you want your presets to only apply to a single catalog, select **Edit | Preferences** (on a PC) or **Lightroom | Preferences** (on a Mac). On the Presets tab, select the **Store Presets With This Catalog** checkbox, as shown next.

When you select this checkbox, all your custom presets will disappear. You can make them reappear by clearing the checkbox, or by re-importing them into your current catalog. With the checkbox selected, each catalog can store different presets.

SUMMARY

Lightroom gives you detailed control over the changes you make to pictures. The History panel lets you browse your changes and go back in time, so you never have to worry about making a mistake. You can use Before & After to instantly see the effect of your changes, either from the beginning of your editing or from any step in your history. Snapshots store a group of settings for a single picture, and presets store settings that can be applied to any picture. You can download free presets or share yours with your friends to speed up editing and for creative inspiration.

13
Customizing Lightroom

To view the videos that accompany this chapter, scan the QR code or follow the link:

SDP.io/LR5Ch13

Lightroom's default settings work for many photographers. If you spend more than a few minutes a day in Lightroom, it might be worth your time to customize Lightroom. Not only can it make you more efficient, but it can make Lightroom more fun to use.

The sections that follow each give a different tip on customizing Lightroom's appearance.

CHANGING THE IDENTITY PLATE

The identity plate is that useless space in the upper-left corner that probably shows Adobe Lightroom 5 or Lightroom Mobile.

You probably don't need to be reminded that you're using Lightroom, so you might as well customize it with your name or the name of your business. Hover your cursor over the identity plate and then click the triangle. Then, select **Change Identity Plate**, as shown next.

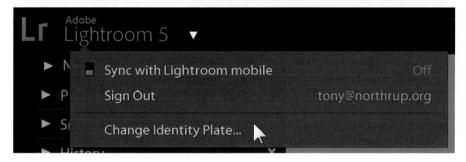

In the Identity Plate Editor, click the list and select **Personalized**.

HOW TO CREATE A TEXT IDENTITY PLATE

From the Identity Plate Editor, choose between text or a graphical logo. For text, select **Use A Styled Text Identity Plate**, choose a font, and type your name. If you change the font and don't see a difference, press **Ctrl+A** (on a PC) or **Cmd+A** (on a Mac) to select your typing.

How to Create a Graphical Identity Plate

If you have a graphical logo, you can use it as your identity plate. For best results, edit your logo so that it's 45 pixels high, and as wide as you can make it. If you use a bigger logo, you'll get strange results. If you use a larger text size (as described in Chapter 14) you'll need a taller logo. You'd think an app like Lightroom could intelligently resize your logo, but it can't.

From the Identity Plate Editor, select **Use A Graphical Identity** plate and then click Locate File.

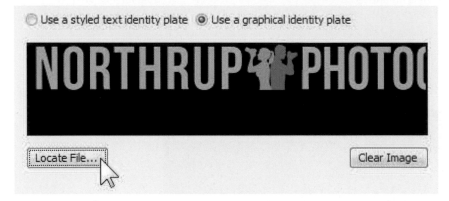

Changing the Module Headers

The module headers are the Library, Develop, Map, Book, Slideshow, Print, and Web links at the top right. The default look is fine, but I prefer to increase the size of the text to make it easier to see and click.

You also use the Identity Plate Editor (**Edit | Identity Plate Setup** on a PC or **Lightroom | Identity Plate Setup** on a Mac) to modify the look of the module headers. Click **Show Details** in the lower-left corner of the Identity Plate Editor to reveal the tools to edit the module headers. I have no idea why Adobe hid the customization options here.

Now, you can adjust the font type, style, and size. You can also set two colors: the left color (orange, in the following example) is the color of the currently selected module, and the right color (blue in the following example) is used for all other modules. You can see that I chose to match the colors to those used in our logo.

The quickest way to browse the different fonts is to click the font type (Magneto, in the previous example) and then move the cursor up and down, previewing the fonts as you go. Choose a size that fills the space; the same size won't work for every font or every monitor, but 35-45 usually looks best on a big screen. Pick something crazy; photography is supposed to be fun!

HIDING MODULES

If you never use one of the modules, it's easy to hide it. Simply right-click any module header and clear the associated checkbox, as shown next. This is a great way to hide the Web module so you don't have to be reminded of how bad the Internet was in 2001.

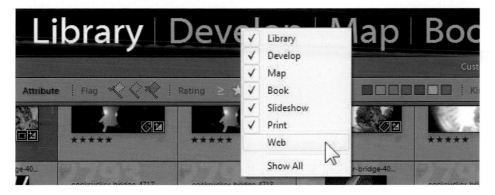

HIDING PANELS

If there are panels you never use, hide them to de-clutter your desktop. Right-click any panel heading, and then clear the checkbox associated with your least favorite panel. Just don't do this if you share Lightroom with another user; most people don't know about the menu.

You can also select **Solo Mode** from this menu. Solo Mode shows only one panel at a time, automatically hiding all other panels when you open a panel.

REPLACE THE SPLASH SCREEN

Lightroom shows a pointless splash screen while it loads (as shown next). You'll never have to see it again.

You can replace it with your own image:

1. Open **Edit | Preferences** (on a PC) or **Lightroom | Preferences** (on a Mac).

2. On the Presets tab, click **Show Lightroom Presets** Folder. We're not actually changing preferences; this is just the easiest way to find the folder. Make note of the path to your Lightroom folder and copy it to the clipboard to make the next steps easier.

3. Close the Preferences window. You don't need to make any changes.

4. Select one or more pictures that you want to be your new splash screen.

5. Select **File | Export**. Configure the Export One File dialog, as shown next, to save the file to a subfolder named Splash Screen of the Lightroom folder you identified in step 2. Click **Export**.

The next time you start Lightroom, you'll see your picture instead! If you export multiple pictures to the new Splash Screen folder, Lightroom will alternate between them.

REPLACE THE END MARKS

End Marks are those decorations at the bottom of the panels, as shown next. They just let you know that you can't scroll further. If you're the type to customize, you can replace them with your own graphics.

First, make a graphic to replace the end mark. As with graphical identity plates, Lightroom isn't smart enough to resize your image, so you'll need to manually resize a logo or graphic to 250-pixels wide or smaller. It can be as tall as you want; you'll be able to scroll to see a tall graphic, but in practice you'll only see the first 200 pixels or so.

For best results, create a PNG file with transparency. Transparent parts of your image will show Lightroom's background color. If the colors in your graphic are too bright, they'll become a visual distraction. I found that setting opacity for the entire PNG file to 25% made the graphics subtle enough to not be annoying.

On the End Marks option, select None to hide them. You can also use your own custom flourishes by following these steps:

1. Right-click the space below the panels (shown in the next figure), select **Panel End Mark**, and then select **Go To Panel End Marks Folder**. Lightroom opens a folder.

3. Copy your own images into that folder.

3. Back in Lightroom, scroll a side panel down to the bottom. Then, right-click the area below panel, as shown next, and select the image you want to use.

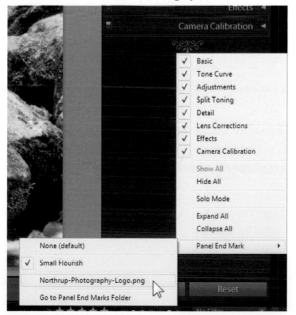

Lightroom will now show your graphic any time there's visible space below a panel, as shown next. You can see that I used Photoshop's sketch filter to convert my logo graphic into something more subtle, so that it wouldn't steal attention away from the more important controls.

CUSTOMIZING THE TOOLBARS

Both the Library and Develop module have a toolbar at the bottom of the window (above the filmstrip). The toolbar can be completely customized—just click the triangle on the right side of the toolbar, as shown next.

Most of the important tools are shown by default, and unnecessary clutter definitely makes photo editing less fun. For example, the Grid Overlay tool in the Develop module is useful for leveling your horizons. However, when you crop and rotate a picture, Lightroom automatically shows you a grid, so you might never need to manually conjure a grid.

Everyone has a different style, however. Check out the toolbar options to see if any of the tools will help you.

CHANGING LIBRARY AND DEVELOP VIEW OPTIONS

You can change many aspects of the Library module to show more, less, or different information about each picture. To modify the appearance of the Library module, make sure you have the Library module selected, and then use the menu to select **View | View Options**.

The Library View Options dialog has two tabs for the Grid View and Loupe View. The options on the Loupe View tab apply only when zoomed in to a picture and when using the Develop module.

GRID VIEW OPTIONS

As shown next, you can configure a great number of different settings to configure Lightroom's appearance when using the Library module's grid view.

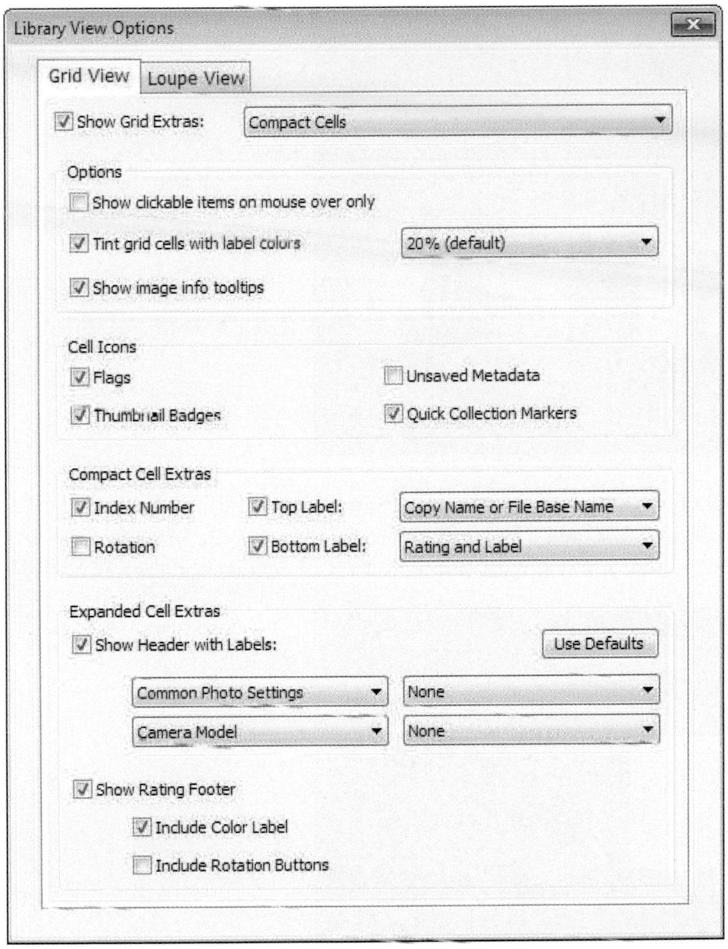

The sections that follow describe each option in more detail.

SHOW GRID EXTRAS

If you have this option selected, Lightroom shows you details of your picture with each thumbnail in grid view. If you clear the option, you'll see just your thumbnails. Clearing the option simplifies the display, but I find the grid extras very useful for finding the exact picture I'm looking for.

When selected, the list allows you to switch between Compact Cells and Expanded Cells. Functionally, they're similar. The Compact Cells, shown next, allow you to fit more thumbnails into the same grid space by cramming the photo information into the space around the thumbnail.

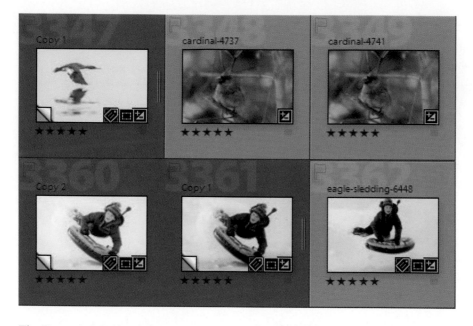

The Expanded Cells, shown next, aren't as cluttered. Instead, the photo information and ratings are shown in separate headers and footers (both of which can be turned off, as described later in this section). However, you can't fit as many thumbnails onto your screen as you can with Compact Cells, because each cell is taller.

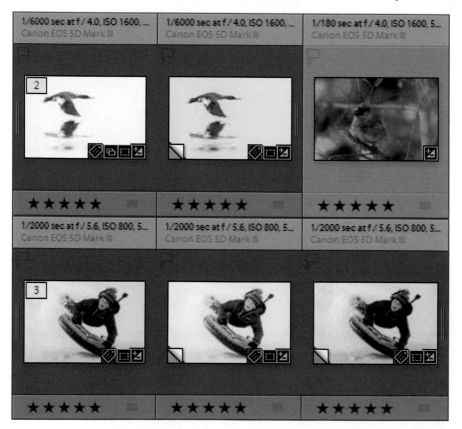

SHOW CLICKABLE ITEMS ON MOUSE OVER ONLY

Lightroom allows you to change attributes of a picture by clicking items on the thumbnails. Which options you see depends on what grid extras you display. For example, if you show ratings and label (as described later in this section) you can click the stars or label to set them for an image.

If you look closely at the next screenshot, you can see the faint outline of a flag in the upper-left corner, five dots (representing five stars) below the thumbnail's left corner, and a grey box (representing the image label/color) below the thumbnail's right corner.

If you have this option selected, those faint outlines are visible only when you move your cursor over a picture. This might require a split-second more time to rate a picture because you'd have to hover your cursor before finding the correct star rating. However, hiding them by default reduces visual clutter.

There are a couple other clickable items that are only shown when you hover your cursor over them, as shown in the next example. Clicking the circle in the upper-right corner adds the image to the quick collection or target collection. If an image is part of a stack, you'll also see a clickable item that allows you to contract the stack.

When this option is selected, you'll only see these clickable items when you hover your mouse over the thumbnail. If you clear the option, they'll be there all the time, cluttering up your thumbnails.

TINT GRID CELLS WITH LABEL COLORS

Lightroom can tint the area behind each thumbnail with the color label. This makes it easier to find pictures with specific labels. This option allows you to control how bright the colors are, and whether Lightroom shows them at all. This next example shows three thumbnails with no label, the yellow label, and the purple label with the maximum 50% tint.

SHOW IMAGE INFO TOOLTIPS

If this option is selected, Lightroom will show some information about a picture when you hover your cursor over the thumbnail. As shown in the next example, Lightroom reveals in a white box below the cursor that I took the photo with my old Canon 10D at ISO 1600.

CELL ICONS

Lightroom can display four types of icons over your thumbnails:

- **Flags** are shown in the upper-left corner, indicating whether you've flagged a picture as a pick or as a reject.

- **Thumbnail Badges** are shown in the lower-right corner over the thumbnail. In the next example, the four badges indicate that the picture has keywords, is in a collection, has been cropped, or has been edited.

- **Unsaved Metadata** is shown in the upper-right corner when applicable, indicating that Lightroom has metadata changes it hasn't yet applied to the original picture.

- **Quick Collection Markers** are the circle in the upper-right corner over the thumbnail, as shown by the cursor in the next example. Click it to add the image to the quick collection or target collection.

COMPACT CELL EXTRAS

Lightroom can show four types of extra information when you are using compact cells:

- **Index Number** shows a big number in the upper-left corner. In the next example, it's 370. Lightroom always starts numbering at 1 for the images in your current grid view, so if you change your filter, the numbers will change. Nonetheless, it's useful when you're showing pictures to a client because they can say, "Let me see picture #370."

- **Rotation.** This shows the counter-clockwise and clockwise rotation icons in the lower corners of the cell. Click these to rotate your pictures. If you don't show this extra, you can rotate pictures by right-clicking them. I don't often need to rotate pictures, so I hide this extra.

- **Top label.** Just above the thumbnail, Lightroom can display information about the picture. In the next example, I'm showing **Common Photo Settings**, which displays as much of the shutter speed, aperture, ISO, and focal length as it can in the width of your thumbnail. The list gives you dozens of different options to choose from.

- **Bottom label**. This is just like the top label, except below the thumbnail. The next example shows the **Rating And Label** option.

EXPANDED CELL EXTRAS

You can also display extra information when using expanded cells:

- **Show Header with Labels**. Select this checkbox to add a header to your thumbnails. You can show at most four pieces of information. Depending on the size and options, you might not be able to see all four pieces of data. The right labels will be hidden completely if the thumbnail is too small. In the next example, I've selected Common Photo Settings and Cropped Dimensions for the top labels, and Camera Model and Title for the bottom labels.

- **Show Rating Footer**. Select this checkbox to add a footer to your thumbnails. The footer always includes the 5-star rating label, and you can choose to add the rotation buttons and color label.

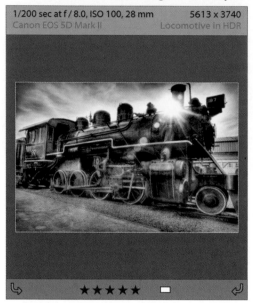

LOUPE VIEW OPTIONS

The Loupe View tab, shown next, configures the extra information you see about each picture when zoomed in using the Library module. The next example shows the camera model, settings, and picture dimensions in the upper-left corner.

You can configure Loupe Info 1 and Loupe Info 2 with different types of information, and then cycle through them by pressing the I key on your keyboard repeatedly. The next figure shows the settings that I prefer.

Here's what the options do:

- **Show Info Overlay**. This chooses which of the two loupe info groups is currently shown. When the options dialog is closed, you can quickly switch between hiding all info and showing each of the two groups by pressing the I key.

- **Loupe Info 1/2**. Use these groups to select which information is displayed.

- **Show briefly when photo changes**. To select this checkbox, first clear the Show Info Overlay checkbox. When you select a picture and have Show Info Overlay turned off, Lightroom will display the info for a few seconds.

- **Show message when loading or rendering photos**. When this checkbox is selected, Lightroom shows a "Loading" message when it can't immediately display a picture.

- **Show frame number when displaying video time**. If you use Lightroom to watch videos, this will show the individual frame numbers, which might be useful for editing (but probably not even then).

- **Play HD video at draft quality**. Lightroom can improve performance by playing high resolution video at lower quality. That's useful if you just want to sample a video, but if you need to assess the quality of your recorded videos, you'll want to clear this checkbox.

SUMMARY

Just like every photographer has a different style, we each use Lightroom slightly differently. While the defaults work for most, customizing Lightroom can make you more efficient. The more efficient your post-processing, the sooner you can be shooting again.

14
Changing Preferences

To view the videos that accompany this chapter, scan the QR code or follow the link:

SDP.io/LR5Ch14

Lightroom has a variety of options that most users will never need to change. To view them, select **Edit | Preferences** from the menu on a PC, or **Lightroom | Preferences** on a Mac. You can also press **Ctrl+,** (on a PC) or **Cmd+,** (on a Mac).

The names don't properly describe some of the tabs, so here's a quick overview of each:

- **General**. Set import options, check for updates, disable the splash screen, and change the language.
- **Presets**. Configure default settings for new pictures, find the Lightroom folder on your hard disk, make presets specific to each catalog, and restore default settings.
- **External Editing**. Configure the file formats used with external editing apps.
- **File Handling**. Configure more import options, and set the cache limits to save disk space.
- **Interface**. Adjust Lightroom's appearance.
- **Lightroom Mobile**. Configure your Lightroom Mobile account or delete your mobile data.

The sections that follow describe each tab in more detail.

GENERAL

The General tab, shown next, allows you to configure import options and user interface options. The File Handling tab also has some import-related options.

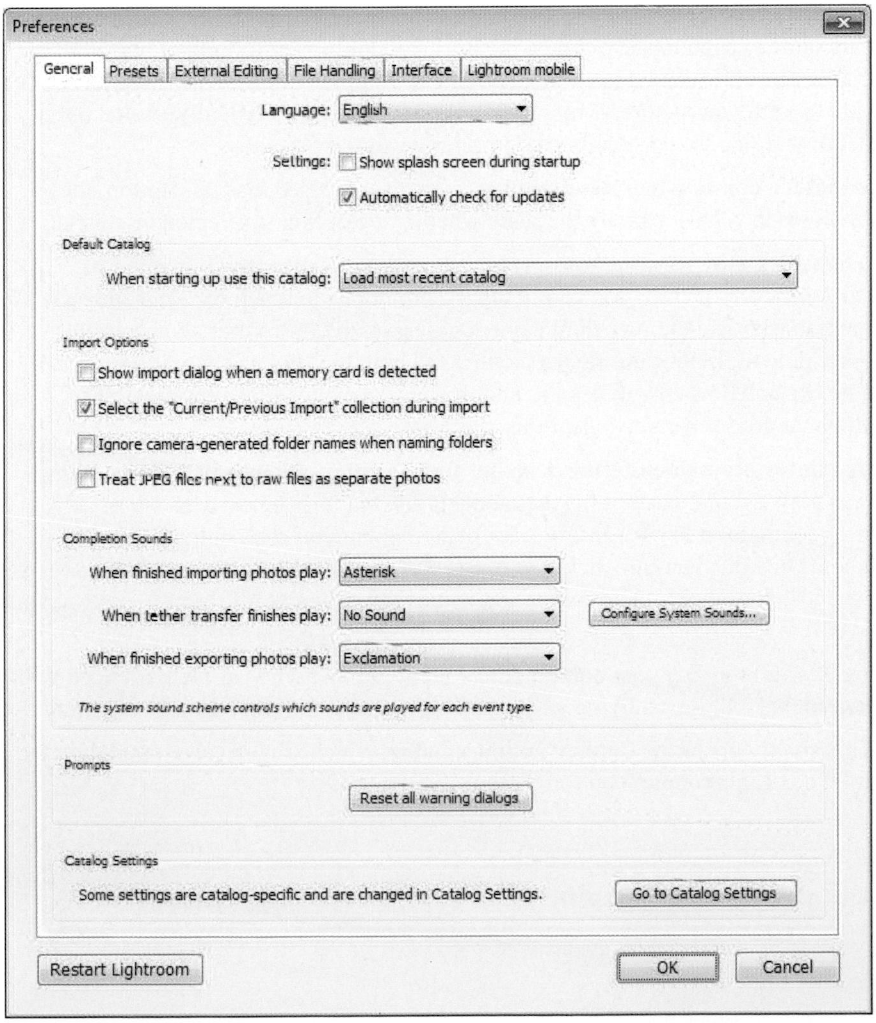

Here's a quick overview of what each setting does:

- **Language**. Choose Lightroom's language (such as English, French, or German).

- **Show splash screen during startup**. The splash screen is the banner Lightroom shows while it's loading. You can clear this checkbox to hide the banner, but it won't speed up loading times. A better choice is to replace the splash screen, as discussed in Chapter 13.

- **Automatically check for updates**. Adobe is constantly improving the latest version of Lightroom to fix bugs and add support for new cameras. There's no reason not to have this checkbox selected because Lightroom won't automatically download or install the update; it will just let you know that the update is available.

- **When starting up use this catalog**. If you only use a single catalog, this setting won't make a difference. If you regularly switch between catalogs, you can choose to have Lightroom load the last catalog you were using (which is generally the most convenient option), load a specific catalog regardless of which you were using the last time you started Lightroom, or prompt you each time it starts to select a catalog.

- **Show import dialog when a memory card is detected**. Selecting this checkbox causes Lightroom to start (if it's not already running) and prompt you to import pictures every time you connect to your computer a memory card, flash drive, camera, smartphone, or anything else with memory. It's a convenient option for many, but if you use other devices or memory cards, having this selected is very annoying.

- **Select the Current/Previous Import Collection during import**. After you start an import, Lightroom (by default) selects the Current Import (also known as Previous Import) collection so you can see your new pictures. Most of us are quite excited to see our new pictures and don't want to look at our old pictures, so that default makes sense. However, it's a bit annoying if you load multiple memory cards, or if you are regularly working on different pictures while importing new pictures. Basically, casual users should leave this selected, but working professionals might prefer turning it off.

- **Ignore camera-generated folder names when naming files**. I've never noticed that this option does anything at all; Lightroom seems to behave exactly the same whether the option is selected or cleared.

- **Treat JPEG files next to raw files as separate photos**. If you shoot JPG+Raw with your camera, your camera creates two files for every picture you take. This is a common approach for Fujifilm photographers and other people who like the way their camera processes JPG files, but also want the option to edit the raw file. By default, Lightroom stacks pictures, so both the JPG and raw file show up as a single photo that can be expanded when you need it, and that option makes sense for most of us. If you'd rather Lightroom show the files separately, clear this checkbox.

- **When finished importing photos play, when tether transfer finishes play, when finished exporting photos play**. Lightroom can play a sound when it finishes some tasks, but it isn't nearly as smart as you might hope. You can't easily configure it to play an MP3, for example, but you can configure it to play one of several system sounds. On a PC, you can click **Configure System Sounds** to change those sounds. If you do change system sounds, they'll change for other applications, too, and you'll have to have a .wav file.

- **Reset all warning dialogs**. If you're giving your computer to a photographer who isn't experienced with Lightroom, click this button to make those annoying warnings reappear the first time they do anything.

- **Go to catalog settings**. This simply opens the Catalog Settings dialog, which you can also open directly from the **Edit** menu (on a PC) or **Lightroom** menu (on a Mac).

PRESETS

The Presets tab, shown next, does more than configure presets; it also configures default settings for newly imported pictures.

Preferences

General | Presets | External Editing | File Handling | Interface | Lightroom mobile

Default Develop Settings

- ☐ Apply auto tone adjustments
- ☑ Apply auto mix when first converting to black and white
- ☐ Make defaults specific to camera serial number
- ☐ Make defaults specific to camera ISO setting

[Reset all default Develop settings]

Location

☐ Store presets with this catalog   [Show Lightroom Presets Folder...]

Lightroom Defaults

[Restore Export Presets]   [Restore Keyword Set Presets]

[Restore Filename Templates]   [Restore Text Templates]

[Restore Library Filter Presets]   [Restore Color Label Presets]

[Restore Auto Layout Presets]   [Restore Text Style Presets]

[Restore Local Adjustment Presets]

[Restart Lightroom]   [OK]   [Cancel]

Here's a description of each option:

- **Apply auto tone adjustments**. Select this and Lightroom will automatically adjust the brightness and contrast of your pictures. This is useful only for absolute beginner photographers.

- **Apply auto-mix when first converting to black-and-white**. If this checkbox is selected, Lightroom converts color images to black-and-white using an auto-mix that makes the blues and greens a bit brighter than the reds and yellows. I prefer the auto-mix; it looks more natural. If you'd rather have all colors weighted equally, clear this checkbox.

- **Make defaults specific to camera serial number**. If you change the default develop settings and this checkbox is not selected, Lightroom applies those settings to all pictures that you import. If you select this checkbox, Lightroom can save different defaults for every camera that you use. If you're correcting camera-specific problems, select this checkbox. Otherwise, leave it cleared.

- **Make defaults specific to camera ISO settings**. Similar to the previous checkbox, Lightroom can configure different defaults for different ISO settings. Select this checkbox if, for example, you want to raise the default noise reduction for high ISO images. I prefer to handle that on an image-by-image basis, so I leave this checkbox cleared.

- **Restore all default Develop settings**. If Lightroom seems to be doing strange things to newly imported pictures, it's probably because you've accidentally changed the default develop settings. Click this button to restore the original defaults.

- **Store presets with this catalog**. By default, Lightroom shows you the same set of presets for all catalogs. That's certainly my preference, but if you'd rather import different presets into every catalog, you can clear this checkbox.

- **Show Lightroom Presets Folder**. Lightroom buries user files in a very difficult to find folder on your hard drive. If you want to browse your files, or drag in new templates, click this button.

- **Lightroom Defaults**. The buttons in this section allow you to reset different aspects of Lightroom back to the defaults.

EXTERNAL EDITING

Most serious photographers use Lightroom to organize their pictures and to perform light editing, but then open other apps (such as Photoshop CC) to make more serious edits. You can use the External Editing tab of the Preferences dialog to configure external tools.

The default settings are good enough for most photographers. If you use Photoshop, Lightroom should automatically detect the app and allow you to transfer a picture into Photoshop with a right-click, as shown next.

Lightroom gives you a few options for how it passes files to Photoshop:

- **File Format**. If you edit a raw or JPG file in Photoshop, Lightroom will send it to Photoshop in its native format. If you make changes in Photoshop and save it, Photoshop needs to save the file in a new format: either PSD or TIFF. Your choice won't change the image quality or editing experience. TIFF, however, is more universally compatible if you use other image editing apps.

- **Color Space**. The default, ProPhoto RGB, is the right choice for most of us. If you need a different color space for compatibility with a printer or your designer, you can choose that instead.

- **Bit Depth**. As with **File Format**, this configures how Photoshop passes photos back into Lightroom, rather than how Lightroom saves files for Photoshop. The bit depth configures how many different gradients of red, blue, and green are available in a saved file. 16-bit takes more space, but 8-bit degrades your image quality. The default of 16-bit is right for almost everyone.

- **Resolution**. This field is meaningless for most of us. It sets the images DPI (dots per inch) or PPI (pixels per inch) as defined in the image's metadata when Photoshop saves an edited file. I can't think of a single modern application that uses that data; obviously, you can print or display your images at any DPI regardless of this value. If you change this field, it will not change your images in any way except to modify this unused setting in the metadata.

- **Compression**. If you use the TIFF format, you can enable lossless compression. Lossless compression means that it saves space, but doesn't degrade image quality. The ZIP option is the most efficient.

Lightroom also allows you to manually configure another external editor. Most of the values are the same as for configuring Photoshop, but the Photoshop settings are applied when Photoshop passes the image back to Lightroom. These external settings are applied when Lightroom passes the image to your external editor.

The most important differences from the previous settings are:

- **Preset**. You can configure multiple different external apps and then switch between them when needed.

- **Application**. Click **Choose** to select the application you want to use.

- **File Format**. In addition to PSD and TIFF, you can export a JPEG file. If you choose JPEG, be aware that the image quality will get slightly worse every time you edit it.

- **Color Space**. You probably won't need to change this, but if your external app supports ProPhoto RGB, I suggest you use that option instead.

- **Compression**. If you use the TIFF format, LZW or ZIP compression will reduce file sizes without reducing image quality. Most newer apps support either, and ZIP is more efficient, so choose ZIP when possible. If your external editor can't read the files from Lightroom, try switching the compression to LZW.

This dialog also provides two options that apply to both Photoshop and your external editor:

- **Stack with Original**. By default, Lightroom stacks edited files with the original, so they are always grouped together. I love stacking, but if you find it confusing, you can clear this checkbox.

- **Template**. Lightroom adds a -Edit to your filename when editing in an external app. You can choose a different template to use another naming strategy.

FILE HANDLING

The File Handling tab of the Preferences dialog gives you additional options for importing photos (which you'll probably never need to change) and configuring the size of Lightroom's cache (which you might adjust to improve performance or save disk space).

Here's a description of the Import DNG creation options, which only apply if you use raw files and convert them to the digital negative (DNG) format. I do recommend DNG conversion, but you generally don't need to change these settings:

- **File extension**. Choose whether you want your file extension to be .dng or .DNG. This won't matter a bit to most of us, even if you're in the very small minority who work with case-sensitive file systems.

- **Compatibility**. Camera Raw is free Adobe software for processing raw files and converting raw files to DNG. Adobe updates it when new cameras are released. You can get the latest version at *sdp.io/cameraraw*. As long as you install the latest version of Camera Raw on every computer that might use your DNG files, you can set this to the latest version.

- **JPEG Preview**. DNG files keep a JPG preview inside the file itself by default. This can make browsing your pictures faster, but it takes a little more disk space. I prefer to have full-size JPG previews, but you can use medium size or eliminate the previews completely to reduce disk space usage and performance.

- **Embed Fast Load Data**. As the name implies, fast load data can improve your performance. Selecting this checkbox causes Lightroom to store a type of preview in the DNG file, increasing file size and performance.

- **Embed Original Raw File**. If you select this option, Lightroom will convert the raw image to DNG but then embed the original raw file. If you don't quite trust the DNG conversion to keep all your camera's original data, select this check box. Your disk space usage will more than double, however.

The next section, Reading Metadata, will only matter when you're importing pictures that already have custom keywords in their metadata, for example, if you're transferring pictures from another app into Lightroom. If you import pictures and find that Lightroom treats multiple keywords as a single keyword, select either or both of these checkboxes. For example, you might import pictures and see that Lightroom considers, "portrait/wedding/cat" as a single keyword instead of three keywords. Selecting **Treat '/' as a keyword separator** and re importing your pictures will cause Lightroom to handle them as separate keywords.

The File Name Generation section has three options, and you'll probably never need to change any of the defaults:

- **Treat the following characters as illegal**. Different computer operating systems, such as Windows, Mac, and Linux (and their file systems), have different rules for file naming. Lightroom is smart about this and replaces the most common illegal characters with a dash (-) by default. If you plan to access your pictures from a different file system that has more illegal characters, you can reduce problems by selecting the longer list of illegal characters.

- **Replace illegal file name characters with**. Choose the character that you want Lightroom to use when it finds an illegal character in a filename. The dash (-) is fine for most.

- **When a file name has a space**. Every modern file system supports spaces in the filenames, but some older file systems do not support spaces. Additionally, accessing files from a command prompt requires surrounding filenames with spaces using quotes ("). That's not something many photographers will ever have to deal with, but if spaces in filenames are a problem for you, change this option to have Lightroom replace them with an underscore (_) or a dash (-).

Finally, there are two sections for configuring the caching of raw images and videos:

- **Camera Raw Cache Settings**. It can take a few seconds for Lightroom to convert a raw file into a picture you can view or edit. To save time when accessing the same file repeatedly, Lightroom stores a copy of processing data in the folder shown at Location. Click **Choose** to change that; I suggest picking the fastest disk on your computer if you have multiple drives. Internal drives are usually better than external drives, and if you have a Solid State Drive (SSD), that's the best choice. Naturally, this cache takes up disk space. If you're low on disk space, you can reduce the **Maximum Size**. If you want to improve performance and you have plenty of disk space, you can increase the **Maximum Size**. Click

Purge Cache to clear the cache and temporarily recover your disk space, but Lightroom will fill it up as you access raw files.

■ **Video Cache Settings**. Lightroom also caches data when viewing video files. You should definitely leave the **Limit Video Cache Size** checkbox selected; otherwise, Lightroom will use as much space as it might need, potentially filling your disk. Lightroom uses the Camera Raw Cache location for the video cache, too.

INTERFACE

The Interface tab (shown next) allows you to tweak Lightroom's look, how you enter keywords, and what Lightroom shows on the filmstrip.

The sections that follow describe each of the different option groups.

PANELS

The Panels section has two options: **End Marks** and **Font Size**. The End Marks option provides access to the same tools described in the "Replace the End Marks" section in the Customizing Lightroom chapter.

Though the Automatic setting works for most, you can change the Font Size option if you have a hard time reading the text on your screen. This is especially helpful if you have a newer high-resolution display, such as Apple's Retina displays. If you do change it, you must restart Lightroom to see the effect; click the Restart Lightroom button in the lower-left corner.

LIGHTS OUT

Lights Out mode (activated by pressing the **L** key) eliminates all distractions and allows you to focus on your picture. Unlike full screen mode, Lights Out mode allows you to make edits. These options allow you to change the screen color and how much the panels are dimmed.

BACKGROUND

These options change the background behind your picture. You can see the effect as you make the changes, so flip through them to find your favorite settings. The pinstriping texture is so fine that you might not even notice it on some monitors.

The secondary window appears if you have a second monitor. You can press **F11** to make it appear, or click the '**2**' monitor icon on the Library toolbar.

KEYWORD ENTRY

By default, you can type keywords in the Keywording panel of the Library module and separate them with commas. So, if you type **Chelsea Northrup, Killarney, Ireland**, Lightroom will treat that as three keywords: "Chelsea Northrup," "Killarney," and "Ireland." Notice that it treated "Chelsea Northrup" as a single keyword, even though it has a space in it.

If you change the Separate Keywords Using list to Spaces, it would treat "Chelsea Northrup" as two separate keywords: "Chelsea" and "Northrup." If you select Spaces and need to enter a space in a keyword, you would have to type quotes around the keyword. I can't imagine why you'd want to change that option.

The **Auto-complete text in Keyword Tags field** controls whether Lightroom suggests existing keywords as you're typing. I find that auto-complete improves consistency and speeds data entry. If you find it annoying, clear that checkbox to turn it off.

FILMSTRIP

The Filmstrip section of the Interface tab has five options:

- **Show ratings and picks**. This shows a flag in the upper-left corner above images that you've marked as a pick, and shows the number of stars in the lower-right corner below images that are rated.

- **Show badges**. This shows a series of icons on top of the lower-right corner of an image, indicating whether a photo has keywords, is in a collection, or has been edited. In the next screenshot, you can see the badges just above the cursor.

- **Show stack counts**. This shows whether an image is part of a stack. The next screenshot shows "2" on the second picture and "2 of 2" on the third picture to indicate that those images are part of a stack.

- **Show photos in navigator on mouse-over**. When you hover your cursor over an image, Lightroom shows it in the Navigator panel in the upper-left corner. Clear this checkbox if you'd rather always see your currently selected image in Navigator.

- **Show photo info tooltips**. If you hover your cursor over an image and leave it there, Lightroom will show you some information about the photo, such as the ISO and camera model used.

This next screenshot shows the filmstrip with all options turned on (though not all options apply to all the images).

TWEAKS

The Tweaks section of the Interface tab has two options:

- **Zoom clicked point to center**. When you click a picture to zoom in, Lightroom (by default) zooms in exactly where you clicked. If you'd rather zoom into the center regardless of where you click, you can clear this checkbox.

- **Use system preference for font smoothing**. Operating systems, such as Windows and Mac OS, manipulate individual pixels to make the smooth curves on letters look even smoother. Lightroom, by default, uses its own algorithm for text, and that works best for most displays. If you have a special display and you've changed the system font smoothing to work better with it, select this checkbox, and then restart Lightroom.

LIGHTROOM MOBILE

The Lightroom Mobile tab, shown next, allows you to join and sign in to Lightroom Mobile (as discussed in another chapter in this book). Click **Delete All Data** to remove any synced pictures.

There's only one option on this tab: **Prevent system sleep during sync**. Most computers are configured to sleep when not in use in order to save power. However, Lightroom can't sync pictures when the computer is sleeping. If you want to prevent sleeping, select the checkbox. This will use more power, but your syncing will finish faster. Once your pictures are synced, your computer will be able to sleep.

SUMMARY

Most people won't ever need to change Lightroom preferences. However, those of us serious about Lightroom can make it much easier and quicker to use by modifying a few settings.

For me, the most important settings are:

- **General tab**: I clear the **Show Import Dialog** checkbox, because I find it annoying when Lightroom opens just because I connected a flash drive.

- **File handling tab**: I reduce the size of the video cache so it doesn't consume too much disk space. I'll often adjust the camera raw cache bigger or smaller depending on how much free disk space I have or move it to a faster drive.

- **Interface tab**: I set a larger font size when I'm connected to a high-resolution display, such as a 4k monitor.

15

Importing, Exporting, & Watermarking Pictures

To view the videos that accompany this chapter, scan the QR code or follow the link:

SDP.io/LR5Ch15

Chapter 1 gave you a quick overview of importing and exporting pictures, allowing you to get started without digging through every complex detail of the Lightroom import and export tools. That overview will be enough for many, but a few of us need more complete control over our photos.

Using the import tool, you can import pictures from your hard disk, a memory card, a camera, or another Lightroom catalog. During the import process, you can weed out pictures that you know you don't want, add metadata such as keywords, rename images, and convert them into the DNG format.

When it's time to share your pictures, you'll want to either export or publish them. Exporting is the more commonly used tool, saving your images to a folder on your hard drive or sending the pictures to a different application. The export tool allows you to watermark your images automatically, adding your name, copyright information, or logo directly to the exported picture.

Lightroom provides a handful of publishing tools that connect directly to social networks such as Facebook and Flickr. Using these publishing tools can be much faster than exporting your pictures and then manually uploading them using your web browser. You can find additional tools for other services, too.

For information about importing existing pictures, refer to Chapter 1. For information about configuring import preferences, refer to Chapter 14.

IMPORTING FROM A MEMORY CARD

If you've selected the **Show import dialog when a memory card is detected** checkbox on the General tab of the Preferences dialog, Lightroom will automatically open the Import dialog when you connect a memory card. Otherwise, connect a memory card or your camera to your computer and select **File | Import Photos And Video** from the menu. You can also press **Ctrl+Shift+I** (on a PC) or **Cmd+Shift+I** (on a Mac).

Usually, Lightroom will automatically select your memory card or camera. If it doesn't, click **Select a source** in the upper-left corner and then select your memory card. If you don't see it on the list, try reconnecting the memory card, connecting the memory card to a different USB port, or turning your camera on.

When importing new pictures from a memory card or camera, I suggest using these options:

- **Source**. Select your memory card, and select the **Eject After Import** checkbox.

- **Copy as DNG**. If you shoot raw files, this will convert them to the more efficient and universal DNG format. Import will take slightly longer than if you use the Copy option, but it will save disk space and can make editing pictures faster. JPG files aren't automatically converted to DNG.

- **New Photos**. Below the copy options, you can select **All Photos**, **New Photos**, or **Destination Folders**. The New Photos option ignores pictures that you've already imported, so if you didn't format your memory card, you won't get two copies of the same picture.

- **File Handling**. I suggest building 1:1 previews to make reviewing your pictures faster. Lightroom can automatically clean up previews to save disk space, as discussed in Chapter 14.

- **File Renaming.** I don't suggest renaming files; if you do, it makes it harder to determine which files from your memory card have already been imported. Instead, leave the **Rename Files** checkbox cleared.

- **Apply During Import.** In this panel, I suggest using the **Develop Settings** list to select a preset that automatically applies lens corrections. For more information, refer to Chapter 12. In the **Metadata** list, click **New** to add custom metadata, such as your copyright information. In the **Keywords** text box, add words that describe any of the pictures you are importing—it's okay if the words don't apply to all pictures because you can always remove keywords, and it's easier to sort through too many results than to search for a keyword and have it return no results.

- **Destination**. Select the folder to store your pictures within. Lightroom automatically creates subfolders organized by date, but you can choose a different technique by changing the **Organize** list. Leave the **Into Subfolder** checkbox cleared unless you want to create a special subfolder within the date subfolder.

This screenshot shows these options in use. Note the lower-left corner, near the cursor, which indicates the total size of the pictures to be imported. Make sure your drive has enough free space. Lightroom shows you the remaining free space on your drive in the Destination panel:

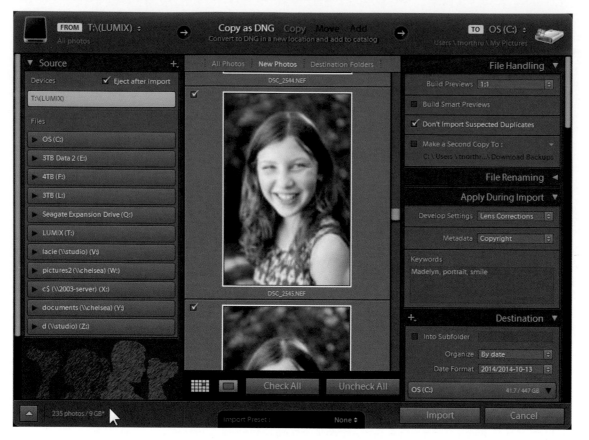

You can reduce the import time by clearing the checkboxes of any photos you don't need to import, such as thumbnails that are completely blurry or dark. If you only want to import a few pictures, click the **Uncheck All** button, and then select only those pictures that you want to import.

At the bottom center of the Import dialog, below the thumbnails, you can create or open an import preset. Import presets simply define all the settings from the Import dialog.

After clicking the Import button, Lightroom begins the import process. You can monitor the progress in the upper-left corner, as shown in the next figure. Importing happens in two phases:

- **Copying and converting pictures**. The speed of this process is mostly based on how fast your memory card is connected to your computer. A USB 3 connection will make this much faster. If you are converting raw images to DNG, that will slow down the import, too.

- **Rendering previews**. The speed of this process is based on the speed of the hard disk you copied the images to and your computer's processor.

You cannot begin importing more pictures until after the first phase is complete. You can import pictures from another memory card once Lightroom begins rendering previews. You can cancel the current phase by clicking the X on the progress bar.

By default, Lightroom selects the **Current Import** collection, also shown in the next figure. If you want to see the rest of your pictures, select the **All Photographs** collection above it. If you don't want Lightroom to automatically switch to this collection, open the Preferences dialog (**Edit** | **Preferences** on a PC or **Lightroom**

| **Preferences** on a Mac), select the **General** tab, and clear the **Select the 'Current/Previous Import' collection during import** checkbox.

IMPORTING FROM ANOTHER CATALOG

If you work with more than one catalog, you can combine pictures into your current catalog. From the menu, select **File | Import from Another Catalog**. Lightroom will import all the pictures and edits into your current catalog. If you create separate catalogs for different clients, it's useful to import the catalogs into your main catalog to make them more easily searchable.

EXPORTING FILES

Before you can use pictures you've edited in Lightroom, you must export them. Typically, you'll export pictures as a JPG file for sharing or printing.

Chapter 1 gave you a brief overview of exporting pictures, and that overview is enough for most users. However, the Export tool is extremely powerful, allowing you to create pictures for a variety of different purposes.

The first step in exporting is to select the pictures that you want to export. You can export a single picture or multiple pictures. Then, from the menu, select **File | Export** or press **Ctrl+Shift+E** (on a PC) or **Cmd+Shift+E** (on a Mac). This opens the Export dialog box, shown next.

The Export dialog is incredibly complex; Adobe crammed every imaginable option into a single window. The sections that follow describe each section in detail.

EXPORT TO

As shown in the next example, the Export To list defaults to your hard drive, and that's the option you'll use most of the time. However, you can also export to email, a CD/DVD drive, an FTP server on the Internet, or a plug-in.

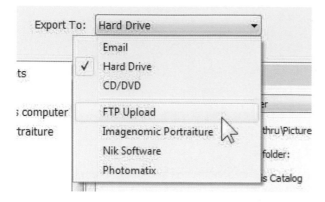

The Email export harkens back to the days when most people used a separate email client application. Today, most people use their web browser to read their email. Usually, when I want to send pictures via email, I'll export them to a folder and then drag them into Gmail in my browser. If you'd like to use the Lightroom email export tool, you'll see the following outdated window.

The FTP Upload is another outdated way to get your pictures on the Internet; for better results, use a custom publishing tool, as described later in this chapter.

The Export To list also shows three plugins that I have installed: Imagenomic Portraiture, Nik Software, and Photomatix. Each of those tools also offers an option when you right-click an image and select **Export**, and I prefer to use the right-click menu.

The sections that follow will assume you're exporting to your hard drive. The options will change if you export to another location.

EXPORT LOCATION

Use this section, as shown in the next example, to choose the folder to store your exported images.

You have several options:

- **Export To**. Typically, you'll select a specific folder and then click the **Choose** button. You can also save the new images in the same folder as the original files (with a different name, of course) or not specify a folder name, which is only useful when creating an export preset (discussed later in this chapter).

- **Put in Subfolder**. I tend to export all the pictures I'm going to share into a single folder. To make it easier for me to find the latest set of pictures, I select this checkbox and then type a unique name that describes the current export. In the previous example, the name was "Cowboy Pictures."

- **Add to this Catalog**. Selecting this checkbox will add your exported images back into your catalog. I don't normally do this, since the exported images are always lower quality than the originals, and it's easy to re-export pictures.

- **Existing Files**. If a picture with the same name already exists in the folder you specify, Lightroom will (by default) ask you whether you want to generate a unique name or overwrite the existing file. Click this list to force Lightroom to automatically overwrite the older image, rename the new image, or skip the export.

FILE NAMING

The File Naming section allows you to rename your files. If you don't select the **Rename To** checkbox, Lightroom will save the exported file with the original filename, but change the extension to .jpg. I usually prefer to keep the original filenames; that makes it easier for me to find the original file if I need to.

If you want Lightroom to automatically rename your files, select the checkbox, then click the list and select **Edit**, as shown next.

Lightroom opens the Filename Template Editor. If you're a computer programmer, you'll feel slightly uneasy with this tool. If you're not a computer programmer, you'll probably hate this tool. In the next example, I've created a filename template that names exported files using the Title of the image (as defined in the Metadata panel of the Library module), provides the resolution, and then numbers them like "043 of 220." For example, it might generate the filename, "Cowboy (7360 x 4912) 002 of 017.jpg".

To use this odd tool, type any custom text in the box near the top. To use image properties in the filename, click in the text box where you want to inser the data. Then, choose the data you want to use from one of the six lists and then click the associated Insert button.

VIDEO

If your export includes video files, and you're not especially picky about your video quality, you can use Lightroom to either copy the original video or transcode the video into a smaller file and resolution. Lower qualities produce smaller files, but they'll be less smooth and sharp.

▼ Video		
☑ Include Video Files:		
Video Format: H.264 ▼	Source:	3840x2160, 29.970 fps
Quality: Max ▼	Target:	1920x1080, 29.970 fps, 22 Mbps
As close to source as possible.		

If you choose to transcode it, I suggest the H.264 option because it's widely accepted. Lightroom won't allow you to export at higher than 1080p.

FILE SETTINGS

If your export includes pictures, use the File Settings section to configure the image format and quality. Most of the time, I use JPEG at 70%, and nobody has complained yet. If I'm making a print, I might bump the quality up to 85%, though many people insist on using the maximum quality for prints. If you want to be sure you don't lose any quality, change the **Image Format** list to TIFF. However, most websites and printing services require JPEG files.

▼ File Settings		
Image Format: JPEG ▼	Quality:	70
Color Space: sRGB ▼	☑ Limit File Size To:	2000 K

Some websites require images to be under a specific size. For example, some sites limit image uploads to 2MB. Each MB is 1,000K, so I entered 2000 in the **Limit File Size To** box to get the maximum image quality within the website's limits.

The default color space is perfect for almost every situation. If your designer or printer asks for another color space, you can change it here using the **Color Space** list.

IMAGE SIZING

This section configures Lightroom to resize your images to specific dimensions. This is rarely necessary; if you're sharing pictures on social media or sending them to a printer, you should simply upload them at full resolution. The next example shows exporting a picture so that neither dimension is longer than 1,000 pixels.

▼ Image Sizing		
☑ Resize to Fit: Long Edge ▼	☑ Don't Enlarge	
1000 pixels ▼	Resolution: 240	pixels per inch ▼

The **Resize To Fit** list gives you several options:

- **Width & Height**. Use this option to specify exactly how wide and tall an image will be. Lightroom won't change the shape of your picture; the values you enter are maximum values, so one dimension can be shorter.

- **Dimensions**. Use this option to specify the image dimensions, regardless of whether an image is portrait (vertical) or landscape (horizontal).

- **Long Edge**. If a website requires images to be smaller than a specific resolution, select this option and specify the number of pixels.

- **Short Edge**. If you want to make the shorter dimension a specific length, select this option.

- **Megapixels**. Select his option if you want the total image size to be a specific number of pixels.

If you're resizing your pictures so that they're not more than a specified resolution, select the **Don't Enlarge** checkbox. If you want your pictures to be exactly specific dimensions, even if they're smaller, clear the checkbox. Enlarging your pictures will make them blurrier.

The **Resolution** box is useful if you are exporting an image for print. Select the printing resolution (often 300 pixels per inch) and then change the **Pixels** list to inches or centimeters.

OUTPUT SHARPENING

If you haven't manually added sharpening to your pictures, the **Output Sharpening** option will do it for you. You get to choose from three different formats: **Screen** (for online use), **Matte Paper** (for paper with flat reflective qualities) and **Glossy Paper** (for shiny prints). You can adjust the amount to your personal taste.

METADATA

The Metadata section (shown next) controls which of your pictures' metadata is saved in the exported file. If you don't have a reason to hide the metadata, you can include all the metadata. If you use titles, captions, or keywords that you don't want the outside world to see, set the Include option to **Copyright & Contact Info Only**.

If you have a camera with GPS and you want to include the location information, clear the **Remove Location Info** checkbox. Some websites, such as Facebook, can read this location information and show a map of where the picture was taken. If you're concerned about privacy, and you don't want the world to know where you are, leave that checkbox selected.

The **Write Keywords as Lightroom Hierarchy** checkbox isn't especially useful. It doesn't change the Keywords metadata of exported pictures, and that's what most applications will read. It does change a separate metadata field, Hierarchical Subject, but few applications read that. It doesn't hurt to select it, but you probably won't see a keyword hierarchy reflected in whichever app you import the pictures into.

WATERMARKING

Lightroom can automatically add a watermark to your pictures when you export them. A watermark is text or graphic overlaid on your pictures. Photographers use watermarks for two main reasons:

- **To protect their images from theft.** While it would violate your automatic copyright on an image, people online can copy your pictures and use them as their own. Adding a watermark can make it more obvious that someone stole your picture.

- **To advertise their services.** If your picture gets shared on a site like *reddit.com*, it's likely that the person sharing it won't credit you as the photographer or link to your website. By adding your name and website as a watermark on the picture, it can't be accidentally separated from the picture.

I personally do not use watermarks. I'd prefer nobody steal my pictures, but if they do, I know they likely wouldn't have been willing to pay for the pictures anyway, so I haven't lost money. Watermarks take away from the beauty of the original image, which overcomes any urge I might have to take credit for my pictures.

To create a watermark, select the **Watermark** checkbox and then select **Edit Watermarks** from the list, as shown next. I suggest exporting a variety of different types of pictures so that you can test your watermark to see how it looks with different images.

This opens the Watermark Editor, as shown next. Use the **Watermark Style** option in the upper-right corner to choose between **Text** and **Graphic** options. Use **Graphic** if you have a logo that you want to use as the watermark. For best results, use a PNG file with transparency.

If you use the **Text** option, type the text you want overlaid in the box in the lower-left corner. If you want to create the copyright symbol (©) and you have a full-sized keyboard, hold down the **Alt** key (on a PC) and type **0169** on the keypad. If that doesn't work, search the Internet for "copyright symbol," and copy and paste an example from a web page into the box.

When you finish your watermark, click the list in the upper-left corner to save the watermark using a custom name so you can easily recall it.

TEXT OPTIONS

Use the Text Options panel to set the font and color. Selecting **Left**, **Center**, or **Right** doesn't change where in your picture the text will appear; it just changes how multiple lines of text are balanced together.

The **Shadow** option is very important because without it, light-colored text will be invisible on a light-colored background.

WATERMARK EFFECTS

The Watermark Effects panel provides these options:

- **Opacity**. This option controls how much your picture shows through your text or image. 100 means that the picture doesn't show at all, and 0 means that your watermark is completely invisible.

- **Size**. Use the **Size** option to determine how big the text or image will be. The **Proportional** option is the most useful; drag the slider to set the size, and Lightroom will always keep the size of the watermark relative to your image size, even if an image has a different resolution.

- **Inset**. The **Inset** sliders adjust how far away from the edges of the picture your watermark is.

- **Anchor**. Use the **Anchor** options to select where Lightroom will place the watermark.

POST-PROCESSING

By default, Lightroom exports your files and then returns you to whatever you were doing. You can use the Post-Processing panel, shown next, to select an application to open your pictures in when it's done.

If you want to open your pictures in an application that's not on the list, select **Open in Other Application** and then click **Choose**. Alternatively, you can select **Go to Export Actions Folder Now** and then copy a shortcut or an application file into the **Export Actions** folder. The next time you click the **After Export** list, the application you copied into the folder will appear in the list.

PRESET

The Preset panel on the left allows you to save and organize your presets. Click the **Add** button to save your current settings as a new preset, or click **Remove** to delete a preset. Select any preset to use those settings.

You can access presets directly from the menu: **File | Export with Preset**.

EXPORTING USING THE SAME SETTINGS

I often need to export a second pictures to the same location, with the same settings. Simply select the picture and then press **Ctrl+Alt+Shift+E** (on a PC) or **Cmd+Opt+Shift+E** (on a Mac).

EDITING PICTURES IN PHOTOSHOP

Lightroom and Photoshop work together wonderfully. Most photographers organize their pictures in Lightroom and perform basic editing. When a picture needs more serious editing (such as warping, masking, adding layers, and combining different photos) the photographer brings the picture into Photoshop. Closing and saving the file imports the edited photo back into Lightroom and, by default, adds it to a virtual stack with the original image.

If you have Photoshop installed, right-click a picture, select **Edit In**, and then select **Edit in Adobe Photoshop** (shown next).

Depending on the type of picture you selected, Lightroom might prompt you with up to three choices:

- **Edit a Copy with Lightroom Adjustments**. This creates a new file and saves any changes you've made in Lightroom. Choose this only if you're confident you'll never want to undo any of your Lightroom changes after editing your picture in Photoshop; those changes are committed forever in the new copy once you edit it in Photoshop. If you need to change a Lightroom setting that you made before the edit, you'll need to return to the original picture to make the change, and then re-edit it in Photoshop.

- **Edit a Copy**. This creates a copy of the original file without showing any of your Lightroom edits. After you edit the picture in Photoshop, you can copy-and-paste your Lightroom settings onto the edited picture. This should be the default, because it allows you to change your Lightroom settings after editing the picture in Photoshop.

- **Edit Original**. If the file is anything but a raw file, this opens the picture in Photoshop without making a copy. This means that any changes will be destructive, because you're editing your original picture. You should only choose this option when you made a copy of the original earlier, for example, when bringing a picture back into Photoshop for a second round of editing.

You won't usually see that prompt when editing a raw file, because you can directly edit raw files in Photoshop, and Photoshop will make a copy when you save your changes. Photoshop will never modify your original raw file.

After editing your file in Photoshop, simply close it. Photoshop will prompt you to save it as a new file with a .tif file extension (by default). Click **Yes**, and Lightroom will automatically import the new picture and add it to a virtual stack with your original photo.

If you don't see the editing picture in Lightroom, it's probably because your filter or collection selection isn't displaying it. Keep your original picture selected, clear your filters (**Ctrl+L** on a PC or **Cmd+L** on a Mac), and select **All Photographs** in the Catalog pane.

You can change how Lightroom exports pictures to Photoshop. For more information, refer to Chapter 14.

Tip: Lightroom also gives you the option to open a photo as a smart object. Smart objects are supposed to maintain a link between your edited picture and the original picture. In practice, however, Photoshop simply saves the smart object as a new picture, and no link is maintained with the original image.

OPENING PICTURES AS LAYERS IN PHOTOSHOP

Photoshop can add multiple pictures together as layers, allowing you to blend the sky from an underexposed picture with the foreground of an overexposed picture. Select all the pictures you want to open together in Photoshop by **Ctrl-clicking** or **Cmd-clicking** them. Then, right-click one picture, select **Edit In**, and then select **Open as Layers in Photoshop**.

The next example demonstrates opening two pictures as layers in Photoshop. As you can see from the thumbnails, the pictures are bracketed. The picture on the left is overexposed, showing detail in the foreground. The picture on the right is underexposed, showing detail in the cloudy sky. Using layers in Photoshop, we can blend them together to use the best parts of each photo.

The next example shows the same two pictures blended together as two layers in Photoshop. You can see the Layers panel in the lower-right corner. The black parts of the layer mask (next to the thumbnail of the top picture, near the cursor) hides the overexposed parts of that picture, allowing me to show the distant mountains and sky from the underexposed photo.

Lightroom provides two other options for opening multiple pictures in Photoshop:

- **Merge to Panorama in Photoshop**. This uses Photoshop's panorama tool to blend multiple pictures of different parts of a scene together into one big picture. For more information about shooting panoramas, refer to Chapter 2 of *Stunning Digital Photography*.

- **Merge to HDR Pro in Photoshop**. This uses Photoshop's HDR feature to blend pictures of the same scene with different exposures together into a single picture with greater dynamic range that can show detail in both bright highlights and dark shadows. For more information about shooting HDR, refer to Chapter 11 of *Stunning Digital Photography*.

PUBLISHING PICTURES

You can also use the Publish Services panel in the Library module (shown next) to export pictures. Lightroom includes a handful of publish services built in, and you can download new publish services for just about any social network or online photography service.

The Hard Drive publish service is exactly like using the Export tool, except that you drag pictures to presets and then publish images as a separate step.

There's a tempting button at the bottom of the panel: **Find More Services Online**. This brings you to the Adobe Add-ons page, which promises to be a wonderful outlet for finding photography tools. It doesn't deliver, however.

You'll have better luck simply searching the Internet for "Lightroom publish <*service*>". Though I haven't tested all his plugins, Jeffrey Friedl's site at *http://regex.info/blog/lightroom-goodies/* has several useful publishing tools. For more information about finding and installing plugins, refer to Chapter 24, "Free Plugins & Tools."

Once you have a publish service that you want to use, click the **Set Up** link to configure it. The exact steps are different for every service, but usually you need to provide login information. If you've already set up a publish service and you want to reconfigure it, right-click the service and then click **Edit Settings**.

Clicking **Edit Settings** opens the Lightroom Publishing Manager, shown next. Use the **Add** button to create a new publishing service. As the next figure shows, you can have multiple copies of a single publishing service. For example, I have separate publishing services for our professional Facebook page, my personal page, and Chelsea's personal page.

To publish pictures to a service, drag them from the grid module to the service. This doesn't start the upload or export process, however. Instead, Lightroom adds the photos to the **New Photos to Publish** queue for that service.

Once you've dragged all your pictures to the publish service, select it to view the **New Photos to Publish** queue. Make sure you really do want your pictures to be public and then click the **Publish** button in the upper-right corner, as shown next.

Lightroom now publishes the photos and moves them from **New Photos To Publish** to **Published Photos**. The files will be immediately available on your hard drive or social network.

Depending on the publish service, Lightroom might add options to your right-click menu. For example, if you publish to Facebook, you can right-click a picture and select **Show In Facebook** (to open your published picture in your browser) or **Mark To Republish** (to re-post it to your timeline).

You might be able to remove photos from social networks using your web browser as if you had directly uploaded them. You can also select the publish service and press the **Delete** key to remove them. In this case, Lightroom doesn't immediately remove them, but instead moves them to the **Deleted Photos To Remove** queue. To officially delete them, click the **Publish** button.

Some publish services support deleting your pictures from Lightroom, while others (including Facebook) require you to use your browser. Always double-check your social network after deleting pictures to verify that they're gone.

SUMMARY

Before you can edit pictures, you must import them into Lightroom. Lightroom can import pictures from a variety of different sources, including your local hard disk, a memory card, or a direct camera connection.

During the import process, you should add metadata such as keywords to make your pictures easier to find. If you shoot raw files, you should convert them to the DNG format to improve processing performance and reduce storage requirements.

Before you can show off your pictures, you have to export them out of Lightroom. Lightroom provides a huge array of export options, but most of us will simply export JPG files to a folder on our hard disk. If you frequently publish images to an online service, you can find custom plugins that connect directly to the service.

16
Tethering

To view the videos that accompany this chapter, scan the QR code or follow the link:

SDP.io/LR5Ch16

You can use tethering to connect your camera directly to your computer. A few seconds after you snap a picture, it will appear in Lightroom, allowing you to preview your pictures on a big screen and even automatically apply development presets and keywords.

Lightroom supports two different methods for tethering:

■ **USB**. USB tethering connects a wire from your camera to your computer. It's faster and more reliable than Wi-Fi, but you might trip over the cable. Not all cameras are supported; check *sdp.io/TetherSupport* for a complete list.

■ **Wi-Fi**. Using Wi-Fi, you can transfer pictures to your computer and then configure Lightroom to automatically import them. To get reasonable performance, you usually need to shoot JPG instead of raw. If your camera doesn't have Wi-Fi built-in, you can use a special card to transfer files.

The sections that follow will describe both of these options.

TETHERING WITH USB

If Lightroom supports USB tethering with your camera, and you don't mind having a cable hanging from your camera, USB tethering is ideal.

USING THE USB CABLE

First, you'll need a long USB cable; the cable that came with your camera is probably too short. When buying a USB cable, be sure the connector type matches that used on your camera; not all USB connectors are the same.

Technically, the maximum length of a USB cable is 5 meters (about 16 feet). You can buy longer cables; however, they need to be powered to ensure the signal travels the entire length of the cable. Longer, unpowered cables might or might not work.

Before plugging the USB cable into your computer, tie the USB cable to something unmovable, like the leg of your desk. Otherwise, when you inevitably forget that your camera is plugged into your computer and walk too far away, the cable will pull your computer and possibly break the USB port or the entire computer.

CONNECTING YOUR CAMERA

Once your camera is plugged in, select **File | Tethered Capture | Start Tethered Capture**.

Lightroom opens the Tethered Capture Settings dialog, shown next. The settings are similar to those used when importing pictures, as described in Chapter 15.

If you select the **Segment Photos By Shots** checkbox, Lightroom prompts you to name each set of shots, as shown next. You'll be able to change it as you're shooting to keep track of different shoots within a single session.

Lightroom uses the session name and shot name to create a folder structure, as shown next. It's one of the few times you'll need to use the Folders panel to navigate your pictures.

SHOOTING TETHERED

Now, hopefully, Lightroom detects your camera and shows you the following tool. If you see **No Camera Detected**, make sure your camera is turned on and close any other applications that might try to communicate with your camera. If it still fails, double-check that your camera is supported at *sdp.io/tethersupport*.

If your camera is supported and still isn't working, try a different USB cable or a different USB port on your computer. If you're using a Nikon camera, verify that the USB setup menu is set to MTP/PTP (Media Transfer Protocol/Picture Transfer Protocol) and not Mass Storage.

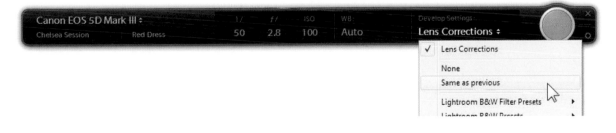

The tethering bar displays the session and shot names, your camera settings, and the develop settings. If you want to automatically apply a preset, select it from the Develop Settings list. You can show or hide the tethering bar by pressing **Ctrl+T** (on a PC) or **Cmd+T** (on a Mac).

You can change the shot name (Red Dress, in the previous example) by clicking it. If you want to change the settings from the Tethered Capture Settings dialog, including the session name, click the tiny gear icon in the lower-right corner.

You can't change the camera's shutter speed, f/stop, ISO, or white balance from Lightroom. Use the buttons and dials on your camera instead. You can, however, click the big button to trigger the shutter.

The picture should quickly appear in Lightroom. If it's slow, try shooting JPG instead of raw. Most cameras don't record pictures to their internal memory card when shooting tethered, but you don't need that because the pictures are already on your computer.

By default, Lightroom selects each new picture as it's taken. If you double-click a picture to view it in the Loupe view, you'll always see the latest picture. If you want to review a single picture without Lightroom automatically jumping to the next picture, use the menu to clear the checkbox at **File | Tethered Capture | Auto Advance Selection**.

MAKING REAL-TIME EDITS

If you want to edit a picture and apply those edits to future pictures, you could create a new preset just for your shoot and then select the preset from the Develop Settings list. Here's an easier way: edit a picture and then select **Same as previous**. Lightroom will automatically apply those settings to all future pictures.

AUTO-IMPORT WITH WI-FI

If your camera isn't supported by Lightroom, or if a USB cable isn't practical, you can connect your camera using Wi-Fi. Lightroom doesn't have Wi-Fi support built-in, but it can automatically import pictures that appear in a folder on your hard drive, and there are several ways to copy pictures from your camera to a folder your PC over Wi-Fi. In order from fastest to slowest, your options are:

- Your camera has Wi-Fi built-in, and the manufacturer provides computer software to connect to the camera and copy pictures to a folder. This includes the Canon 70D, 6D and Nikon D5300, D750. Unfortunately, Wi-Fi cameras from Sony, Olympus, and Panasonic can only transfer images to tablets and smartphones.

- You add a Wi-Fi adapter to your camera and use computer software. Most Canon cameras can use the WFT-E5A, WFT-E6A ($600), or WFT-E7A ($850), and most Nikon cameras use the WU-1a or WU-1b ($43) or WT-5A ($560). However, all of these external adapters are clumsy and wildlife overpriced (even the $43 model).

- The CamRanger ($300) can transfer files to a computer if you have a Canon or Nikon DSLR. For a complete compatibility list, visit *camranger.com/supported-cameras*.

- You use a SD memory card with Wi-Fi. This works with just about any camera, and if your camera uses CF cards, you can use an SD to CF card adapter. The only model that can automatically work with Lightroom is the Eyefi (eyefi.com). Currently, the Transcend Wi-Fi SD card and Toshiba's FlashAir require you to manually copy images from your camera to your computer.

Because every method is different, I can't provide step-by-step instructions for copying pictures from your camera to your PC via Wi-Fi. Once you get pictures automatically copying, from the menu, select **File | Auto Import | Auto Import Settings**, as shown next.

Now, configure the Auto Import Settings, as shown next. For the **Watched Folder**, select the folder your Wi-Fi software is saving pictures into. For the **Destination**, specify a **Subfolder Name** specific to your shoot. Set the **File Naming**, **Develop Settings**, **Metadata**, **Keywords**, and **Initial Previews** as described in Chapter 15.

You can turn auto import on and off using the menus: **File** | **Auto Import** | **Enable Auto Import**.

Lightroom moves pictures from the watched folder to the destination folder as it processes them. Therefore, when you take a picture, your Wi-Fi software will move the picture into the watched folder. There, Lightroom discovers the new picture, moves it into the Destination folder, and processes it. If auto import is working correctly, pictures should only remain in the watched folder for a few seconds.

SUMMARY

Neither USB tethering nor Wi-Fi are ideal, but both are useful in different scenarios. Use USB tethering whenever you can; it's faster and more reliable. Wi-Fi is definitely more convenient, but the slower speeds mean that if you shoot larger raw files, your images might take too long to appear.

17
Playing & Editing Videos

To view the videos that accompany this chapter, scan the QR code or follow the link:

SDP.io/LR5Ch17

Lightroom allows you to organize video in the same ways that you organize your photos. That's genuinely useful because many of us switch between taking still photos and video clips during sports, family events, and travel.

Organizing your video clips isn't much different from organizing your photos (discussed in Chapters 1, 3, and 4). One difference is that cameras don't store as much metadata in the video files, so you won't be able to view the camera model, aperture, shutter speed, or ISO for the video. You'll need to manually add metadata about your videos.

Lightroom includes only the most basic tools for editing your video. You can trim the ends, fix the color, adjust the brightness and contrast, and adjust the vibrance. You can then export the edited video at up to 1080p/30fps.

Using the Slideshow module, you can add a music track, basic titles, and hack together more complex edits. By using virtual copies, you can cut portions from the middle of your video.

You *could* do this, but there are much faster and easier ways to edit your videos. These tools let you easily add titles, crop, adjust sound problems, cut boring sections of video, and cut between multiple video clips:

- **Windows PCs**: Windows Movie Maker (*sdp.io/wmm*) is a free and powerful video editing tool. If you need even more power, Adobe Premiere Elements (*sdp.io/ape*) is about $100.

- **Macs**: iMovie (*sdp.io/imovie*) is $15 and far faster and easier than Lightroom for video editing. If you need even more power, Adobe Premiere Elements (*sdp.io/ape*) is about $100.

Even though it's not always the best tool for the job, this chapter will give a quick overview of Lightroom's limited video editing capabilities.

FINDING VIDEOS

In the Library module, you can use the Attribute filter to show just your videos by selecting the right-most icon, as shown next.

PLAYING VIDEOS

Once you find a video in the Library module, the Play button will let you view it, as shown next.

If playback is slow, you might be able to improve the performance. From the menu, select **View |
View Options**. On the Loupe View tab, select the **Play HD video at draft quality** checkbox, shown
next. You won't see every glorious pixel of your video, but at least playback will be smoother.

Another way to improve playback is to not use Lightroom. Unfortunately, Lightroom doesn't give
you a quick way to open a video in an outside player. Your best option is to right-click the video and
then select **Show in Explorer** (on a PC) or **Show in Finder** (on a Mac). From within Explorer or
Finder, double-click the video to watch it in your default viewer. My favorite video player is VLC
Media Player, available at *www.videolan.org*.

When you play video, Lightroom might use the video cache to increase performance (but use
some of your disk space). You can control how much disk space Lightroom uses from the **File
Handling** tab of the Preferences dialog, shown next. To reclaim the used disk space, click the **Purge
Cache** button. To open the Preferences dialog, select **Edit | Preferences** (on a PC) or **Lightroom |
Preferences** (on a Mac).

SETTING THE THUMBNAIL FOR A VIDEO

You can control frame of the video Lightroom uses for the thumbnail. Pause the video on the frame
you want to use, click the box icon on the control bar, and then select **Set Poster Frame**.

Lightroom also uses the poster frame to display the histogram. If you plan to edit the color and brightness of the video (described later in this chapter), set the poster frame to a meaningful frame in the video so that you won't accidentally under- or over-expose it.

EXTRACTING A STILL PHOTO FROM VIDEO

You can use the same menu to save a frame of the video as a still photo by selecting Capture Frame. Lightroom saves the still photo as a JPG file using the same filename as the video (but adding a -1, -2, etc., to the filename). Lightroom also adds the new still to a virtual stack with the original video. If you don't see the still photo in the Library view, turn off any filters.

If you plan to share your video on YouTube or another video sharing site, extracting a still frame is a great way to create a custom thumbnail. A thumbnail chosen deliberately will get more viewers than whatever random thumbnail the video sharing site chooses for you.

Capturing frames is an amazingly useful technique for action, especially if you have a 4k camera. While 4k video is equivalent to only 8 megapixels, it records at 30 frames per second, which is far faster than I could capture using any still camera. I can then pick the exact frame where the bat hits the baseball or the player's foot meets the soccer ball, extract that frame, and use it as a still photo.

ADJUSTING THE COLOR AND BRIGHTNESS OF VIDEO

You can use the Quick Develop panel to change the white balance, exposure, contrast, whites, blacks, and vibrance of the video, as shown in the next example. Use the **Treatment** list to convert a video to black-and-white.

For some reason, Lightroom doesn't let you edit videos with the Develop module, even though there are many Develop settings that you can apply to videos. Instead, you can edit a picture with the settings you want, save a preset, and apply it to a video clip using the **Saved Preset** list. I suggest extracting a still from the video as described earlier in this chapter, editing it with the Develop module, and then using copy-and-paste to transfer the settings to your video. You can also use the **Auto Sync** tool in the Develop module.

Only the following settings can be applied to video clips:

- White balance
- Exposure
- Contrast
- Whites
- Blacks
- Saturation
- Vibrance
- Tone Curve
- Treatment (color or black-and-white)
- Split Toning
- Process Version

CALIBRATION

That's leaving out many very important edits, including cropping, sharpening, noise reduction, and lens corrections. Each of these are standard features for dedicated video editing applications, such as those described in the introduction of this chapter.

TRIMMING THE ENDS OF VIDEO

You can trim the start and end of a video by clicking the gear icon, shown next.

The gear icon shows thumbnails of your video on a timeline. To trim the ends of your video, drag the marker or play your video until you get to the point where you want your video to start. You can use the buttons to the left and right of the play button (highlighted next) to move backwards or forwards by a single frame.

Once you've found the frame that you want to start your video, drag the left edge of the video to the marker, as shown next. You can also press **Shift+I** to mark the beginning of your video.

Trimming the end of your video is the same process, but the keyboard shortcut is **Shift+O**.

Like everything in Lightroom, your edits are non-destructive. You can always drag the markers back to see the trimmed parts of the video.

ADDING CUTS AND MUSIC TO VIDEO

You often need to trim sections from the middle of a video. For example, if you record your child's soccer game, you wouldn't want to force all your Facebook friends to watch 60 minutes of video to see your kid score. Trim the video to show just the good parts, and you'll get far more people to watch your video. For a soccer game, you might want to show four clips:

- The referee starting the game
- Your child scoring
- The audience applauding
- The team celebrating their win on the sidelines.

If you recorded the entire game in a single clip, this might seem impossible because Lightroom only allows you to trim the start and end of video. There's an easy work around: virtual copies. Make several virtual copies of a video, and in each virtual copy, trim the start and end for that particular video segment. Then, use the Slideshow module to link the virtual copies into a single video.

In this example, you would follow this process to make your final video:

1. In the Library module, adjust the color and brightness of your clip or clips, as described earlier in this chapter.

2. Make three virtual copies of your video, so you have four clips total.

3. Using the first clip, trim the start and end so that it shows only the first segment of your planned video.

4. Trim the start and end of the second video (the first virtual copy) to show only the second clip, your child scoring.

5. Trim the start and ends of the third and fourth videos to show only the audience applauding and the team celebrating.

6. If necessary, use the Grid view in the Library module to drag-and-drop your clips into the correct order. This is a useful technique for adding what videographers call "B-roll"—video clips that help the storytelling but weren't necessarily caught at the same time. In the example soccer game, you could get a clip of the audience applauding at any time during the game and show it after your child scoring. It doesn't make a difference to the viewer that the audience was cheering for a different moment.

7. Select your four video clips and switch to the Slideshow module.

8. In the Template Browser, select the **Lightroom Templates | Crop To Fill** or **Widescreen** template.

9. In the Playback panel, select the **Audio** checkbox and then click **Select Music** to add music behind your video. Music almost always makes a video more entertaining.

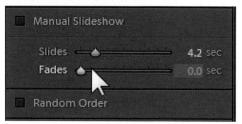

10. Also in the Playback panel, in the Manual Slideshow section, set **Fades** to **0.0 sec** (shown next). While it's okay to have crossfades in video clips, most viewers prefer hard cuts. Also clear the **Random Order** checkbox.

11. Click the **Export Video** button to save the clips as a single, final video.

After cutting your video into clips, you might discover that each clip needs different settings for brightness, color, and contrast. No problem; just jump back into the Library module and edit the individual clips.

For detailed information about using the Slideshow module, refer to Chapter 20.

Sharing Your Video

If you've edited your video using the Slideshow module, you must use the **Export Video** button in the Slideshow module to save your finished product as a file.

You can directly share that video by uploading it to a site like YouTube or Facebook. After exporting the finished product, import it back into Lightroom so you can easily find it later.

You can also use the Publish Services in the Library module to upload individual video clips directly to Facebook or Flickr. In the Publish Services panel, click **Set Up** (if necessary) to connect to your social network. Then, drag your video to the service, shown next.

Now, select the service that you dragged the file into, and click **Publish**, as shown next. Lightroom converts and uploads the video for you, which might take several minutes.

The video will appear on your timeline, as shown next. You might find that the quality has been greatly reduced, however. This might resolve itself in a few minutes (or hours) when the service has time to fully process the video. Depending on the service, you might get better results by exporting a file to your hard drive and using your web browser to upload it.

SUMMARY

Lightroom isn't the greatest tool for editing your videos, but if you just need to make a few cuts, fix color and contrast problems, and add music, Lightroom just might be the fastest way to casually edit a home video and post it to Facebook or Flickr.

18

The Map Module

To view the videos that accompany this
chapter, scan the QR code or follow the link:

SDP.io/LR5Ch18

Of all the ways to organize and find your pictures in Lightroom, the most powerful is the Map module. If you can remember where you were when you took a picture, you can find it on the map. Want to find those pictures you took of the Eiffel tower, but you didn't add any keywords? Just search for Paris and zoom in.

If you find a picture and don't remember exactly where you took it, you can select it and then switch to the Map module to see where it was taken. Did you stumble across a gorgeous view while hiking, and you'd like to return in better light? Lightroom can show you exactly where you were standing. Forget where that restaurant was, but remember that you took a snapshot at the bar? Find that snapshot, and then find your way back.

The Map module works wonderfully when you load pictures that have GPS data. All modern smartphones are capable of recording GPS data, though location information might not be enabled by default. A handful of DSLRs have GPS built-in, including the Canon 6D, Nikon D5300, Sony a65, a77, and a99. You can buy GPS receivers for Canon and Nikon DSLRs, typically costing $100-$400.

For photos that aren't tagged with GPS data, you can either drag them to the map to manually add their location, or use a smartphone app to record your location while taking pictures and then use Lightroom to link your photos and location data.

This chapter covers the Map module, tagging locations, and synchronizing GPS data.

USING THE MAP MODULE

The map module is intuitive to use. Drag the map with your cursor to move it. Scroll in and out with your mouse wheel or use the - and + slider at the bottom-left of the screen. Click the **Map Style** list in the lower-left to change the map from a road map to a satellite map or a terrain map. Use the **Search** field in the upper-right corner to type a location and jump to it.

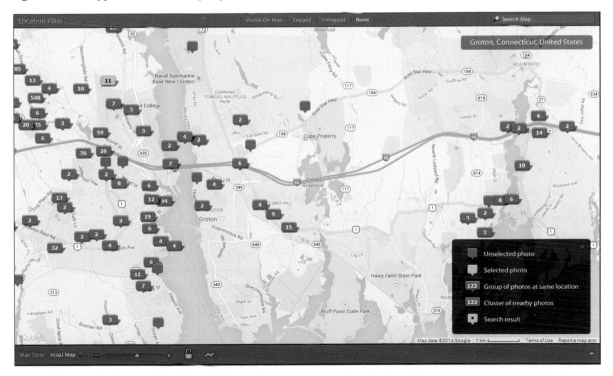

Lightroom shows you your tagged photos; double-click on the orange icons to view the individual photos. Click again on a photo to open it in the Library module for further editing.

FINDING TAGGED PICTURES

The most important use of the Map module is for finding your pictures. Therefore, a common task is to find a set of pictures taken at a location on the map and then move to the Library module to browse the thumbnails in the grid view.

Once you zoom in on the map area containing your pictures, use the **Location Filter** at the top of the map to view the pictures in the Filmstrip. The Filmstrip is the list of pictures at the bottom of the Lightroom window.

Visible On Map (shown next) is the most useful filter; it shows you all the pictures that you've found. **Tagged** shows you only those photos with GPS data (not including photos you've manually tagged) and **Untagged** shows you only those photos you've manually tagged.

After selecting one of the filters, switch back to the Library module (or press **G**). The filtered photos will be selected using the **Map Location** filter, shown next.

Library Filter :		Date
Map Location		
All (3 Map Locations)	1200...	All
Amsterdam	331	▷ 200
Current Map Location	3867	▷ 200
Unknown	1158...	▷ 201

MANUALLY TAGGING LOCATIONS

If some of your pictures don't have GPS information, you can tag a photo with the location simply by dragging it from the filmstrip to a location on the map. You can also drag individual tagged photos when you see the icon shown next; this is useful for adjusting an incorrect location.

SYNCING LOCATION DATA FROM YOUR SMARTPHONE

Very few cameras have GPS data built in. If your camera doesn't, you can use a smartphone to record a GPX data track while you're taking pictures. Your phone will record the time and your location on regular intervals, and then Lightroom will use the Date Captured metadata that your camera records to determine where you were when you took each picture. In other words, run a smartphone app while you take pictures, and Lightroom can show your pictures on a map.

First, set the time on your camera to the time on your smartphone. Make sure you set the time zone on your camera.

Next, find a smartphone app by searching your app store for "GPX Log." There are free apps that do the job; the app I use is simply called "GPX Logger." On my app, I simply hit the Rec button to start recording the location. Make the first picture you take a picture of your GPX app showing the current time. If the time on your camera and in the app don't agree, this will allow you to adjust it later.

Depending on your phone and the app, you might have to leave the app running in the foreground to record data; switching to another app might stop it. Be aware that recording your location uses a great deal of battery power, so plan accordingly. Keep a USB battery charger and cable in your bag so that your GPX app doesn't leave you stranded without a phone.

After you're finished shooting, you'll need to copy the GPX log from your phone to your computer. That's different for every app, so refer to the app instructions.

Once you load your pictures into Lightroom, use the Library module to select the pictures that you want to tag with GPS. Now, switch to the Map module, click the GPS Tracklogs button at the bottom of the screen, and then click Load Tracklog, as shown next. Select your tracklog file.

Now, Lightroom will attempt to find pictures that were taken at the same time as your GPS tracklog. You should verify that the time on your tracklog and the time on your photos match. Click the GPS Tracklog button again and select **Set Time Zone Offset**. Lightroom displays the Offset Time Zone dialog, shown next. Slide the Offset slider until the times on the photos and tracklog overlap.

Click **OK**. You're now ready to sync your photos. Click the GPS Tracklog button one more time, and then **click Auto-Tag Selected Photos**. You should see your pictures appear on the map along your tracklog.

SUMMARY

It's a shame more cameras don't have GPS built in. Those that do typically turn it off by default because it uses a great deal of battery power. Nonetheless, if you're deliberate about it, you can add location information to any picture. It's incredibly rewarding when you do have GPS data, because you can see your travels in an entirely new way.

19
Making a Photo Book

To view the videos that accompany this chapter, scan the QR code or follow the link:

SDP.io/LR5Ch19

Photo books are an amazing way to keep and share your pictures. For $30-$100, you can print a gorgeous, hardcover, full-color book that will make your images look amazing. I create photo books for vacations, weddings, and even birthday parties. They make unforgettable, creative gifts that the receiver will cherish for years.

In this chapter, I'll show you how to use Lightroom's Book module. However, I start the chapter by warning you that the Book module might not be the best choice for your project. For those who choose to use a different book printing service, the tips in this chapter will save you time and help you make the most beautiful book possible.

CREATING A COLLECTION FOR YOUR PHOTO BOOK

First, create a collection with your pictures. I suggest using virtual copies of your pictures to allow you to freely edit the pictures. For example, you might find that differences in the white balance make two pictures placed on the same page look unnatural. Fitting a picture onto a page might require you to use a wider crop than you would on the web. Virtual copies will allow you to edit freely without changing your original pictures.

Most photo book layouts require unusual cropping. It's easy to add crop to pictures in the Book module, but if you want to un-crop pictures, you'll have to switch to the Develop module. To speed the book layout process, you might want to remove the crop from all your pictures.

To remove the crop from all your pictures, select them all (**Ctrl+A** on a PC or **Cmd+A** on a Mac). Switch to the Develop module and turn **Auto Sync on**, as shown next.

Now, click the **Crop** tool and click **Reset**, as the next example shows. Now, all your photos will be returned to their original crop. If you've cropped your pictures and then edited them outside of Lightroom (for example, in Photoshop), you might want to return to the original picture so that you don't lose any cropping potential.

With your new collection selected, switch to the Book module. As shown next, Lightroom's Auto Layout feature (described later in this chapter) fills the pages with your pictures. The top two pictures are your book's back cover (on the left) and front cover (on the right).

CHOOSING A PRINTING SERVICE

The **Book** list in the Book Settings panel allows you to choose between **Blurb**, **PDF**, and **JPEG**. Blurb is a book printing service that has partnered with Adobe, and they're the only service to which you can send your book directly from Lightroom.

That's the biggest weakness with the Book module; you can't send your book to Mixbook (my favorite), Shutterfly, MyCanvas, or any other printing service. Blurb is a great service, but it's not necessarily the best service. You might find better prices, higher quality, and more size options from other book printing services.

In theory, you could select the **PDF** option in the **Book** list and send your PDF to any printing service. In practice, Lightroom doesn't allow you to customize the size of the book, so you're limited to the size options that Blurb uses. Those aren't necessarily the same sizes used by other printing services. Additionally, other printing services might have specific requirements for how pictures are laid out.

In other words, the PDF option isn't a practical way to have a photo book printed with another service. However, all other photo book printing services offer nice tools for laying out your photo book, and some of them work even better than the Book module.

It's nice that you can lay your book out directly in Lightroom, but every book service offers easy-to-use graphical tools. In fact, Blurb's own downloadable tool (BookSmart) is more powerful than Lightroom's Book module.

I had to suggest other options because photo books are expensive and the memories captured within are important. Adobe partnered with Blurb because they're paying Adobe, not because Blurb is always the best choice. For the rest of the chapter, I'll tell you how to use the Book module and give tips that are applicable to any photo book service.

CHOOSING A BOOK SIZE

First, decide on the size of your book. That's the first choice you need to make, because you'll need to arrange and crop your pictures into the page size you select. The **Size** option in the **Book Settings** panel lets you choose from Blurb's sizes:

- Small Square (7x7 in. or 18x18 cm)
- Standard Portrait (8x10 in. or 20x25 cm)
- Standard Landscape (10x8 in. or 25x20 cm)
- Large Landscape (13x11 in. or 33x28 cm)
- Large Square (12x12 in. or 30x30 cm)

If you plan to print full-page pictures, the landscape options will usually be the easiest to layout, because most pictures are taken in landscape format. If you followed my advice in Chapter 1 of Stunning Digital Photography to know your final format and leave room to crop, you might be able to fit your pictures into other formats.

If you typically fill the frame with your subject, you will struggle with full-page pictures because of the necessary cropping and parts of the picture lost in the gutter (the center of the book, where the binding is). That's okay; you can simply print your pictures with their existing crop and leave white space around them.

I've never thought a photo book was too large; the Large Landscape book has far more impact than the Standard Landscape. The Small Square format is very cute, however, and makes a great gift.

CHOOSING PRINTING OPTIONS

You should also use the Book Settings panel to choose options for your cover and paper type. You'll definitely notice the difference between the Standard page type and the Premium and high-end ProLine paper options. The more expensive papers are thicker and help prevent pictures from bleeding through to the opposite side.

If you choose the Premium or ProLine paper options, you have the option between glossy (Premium Lustre or ProLine Pearl Photo) or matte (Premium Matte or ProLine Uncoated). Glossy pages crisply reflect light, whereas matte acts like a diffuser over your picture. Glossy pages are the right choice for most books. If the mood of your book is serious, the matte options might be a better choice.

The **Logo Page** option defaults to **On**. Blurb gives you a discount if you let them print a small advertisement on the last page of your book, a practice that's pretty standard among book printing services. You can turn off the logo page option for an extra fee, but I've never had anyone complain, so I'd rather spend my budget on nicer paper or a better cover.

USING AUTO LAYOUT

Lightroom can automatically place pictures on the pages of your book. The Auto Layout panel offers three options:

- **One Photo Per Page**. My favorite starting point, this fills each page with a picture.
- **Left Blank, Right One Photo**. Leaves the left page blank. This provides the greatest photo quality because it prevents the pictures on the opposite side from bleeding through. You can also use the blank page to add a description of each picture, though the text might be visible in light parts of the opposite side.
- **Left Blank, Right One Photo, Caption**. This adds a small text box below each picture that you can fill in with your own caption. It does not read the caption from the metadata.

Click the list to select a different option, and then click the **Auto Layout** button (shown next). If you want to start with a blank book and place each picture manually, click **Clear Layout**.

If you're making a short book, it's probably easiest to manually place each picture. If you're making a longer book, it might be worth the time to create a custom auto layout preset. By creating a custom auto layout preset, you can use Lightroom to get your book as close as possible to the final layout, and then make page-by-page adjustments as needed.

To create an auto layout preset, click the **Preset** list and then select **Edit**. You'll see the Auto Layout Preset Editor, shown next. Select the default layout for the left and right pages, choose the zoom default, and add text from your picture's metadata.

The **Zoom Photos To** list is important. Fill crops your picture, but fills the entire space provided. Fit doesn't crop your picture, but might leave extra white space around your photo.

ADJUSTING PICTURES

Clicking a picture allows you to drag it to change its position in the frame (shown on the right). Adjust the **Zoom** slider to increase the photo's crop.

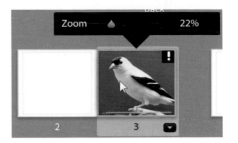

CHANGING PAGE LAYOUTS

The auto layout is just a start; each page can use more than 100 different layouts. Click a page to select it, and then click the down arrow to select a different layout.

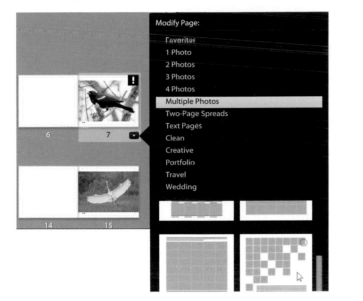

Many of the layouts allow you to show multiple pictures on a single page. Grouping pictures together is a powerful storytelling technique, but it requires you to think about the relationship between the pictures. Don't place unrelated pictures in close proximity.

ADDING PAGES AND PAGE NUMBERS

Use the **Page** panel to add blank pages and to configure page numbers. To add a page, click the list and select the layout, as described in the previous section. Then, click **Add Page**, as the next example shows. To add a blank page, click **Add Blank**.

Most photo books don't need page numbers. If you can imagine needing to tell someone to flip to page 37, then you probably need page numbers. Otherwise, numbers are simply a distraction.

ADDING TEXT

There are three ways to add text to a page:

- Choose a page layout that includes text. Then, type the text in the box provided.
- Select a picture, and then select the **Photo Text** checkbox on the **Text** panel (shown next). Then, select the new text box and type the text.
- Select a page, and then select the **Page Text** checkbox on the **Text** panel. Then, select the new text box and type the text.

Use the Type panel, shown next, to change the size, font, and layout of your text. You don't have to use the same size and font throughout your book, but if you don't, your book might seem unprofessional.

SETTING THE BACKGROUND

Most photo books should have a plain white background; this reduces visual clutter and helps focus the reader's attention on your pictures. You can use the Background panel, shown next, to change the background color or add a graphic. Click the triangle to the right of the **Drop Photo Here** space to select one of Lightroom's tasteful backgrounds.

If you choose to use your own graphic, you'll get the best results if you use a style similar to the built-in graphics: monochromatic and faint, like a proper watermark. Anything bolder will be distracting.

If you choose a different background color, keep in mind that it won't actually change the paper color. Your paper will always be white, but the printer will completely cover it with ink, which might create a different effect.

ADDRESSING RESOLUTION PROBLEMS

If you see an exclamation point in the upper-right corner of one of your pictures, as shown in the next example, Lightroom is telling you that your picture's resolution isn't high enough to look sharp when printed.

You don't need to fix this; you can print it with a low resolution and your picture will just be a bit unsharp. I've never had anyone complain. However, if you're concerned about sharpness, you can fix this in a few ways:

- Change the photo layout so that the picture is printed smaller.
- Change the crop or zoom of the picture so that it's not as heavily cropped.
- Use a smaller book layout.

And remember this moment the next time someone tells you that megapixels don't matter; as a photographer doing print layouts, it's shockingly common to wish you had more pixels to work with.

PRINTING YOUR BOOK

Finally, click the **Send Book to Blurb** button in the lower-right corner. This hands you off to the Blurb website, where you'll create a login and pay for your book.

SUMMARY

Photo books are an amazing gift, even if it's a gift to yourself. I've seen people moved to tears from a book, and the book can last decades. Sure, you could look up your digital images on your computer, but there's simply no replacing a book on the coffee table that you can flip through at any time. Your great-great-grandkids will appreciate the glimpse into your world, too.

20
Slideshows & Time Lapses

To view the videos that accompany this
chapter, scan the QR code or follow the link:

SDP.io/LR5Ch20

For decades, slideshows were how many serious photographers displayed their artwork (especially in Europe, where shooting with slides was more popular than using negatives). Photographers would compile their best slides from a trip or event, load them into a slide projector, pull the big white projection screen from the ceiling, cut the lights, and talk about their pictures one by one.

It was an amazing way to present your pictures. You had everyone's total attention; the room would be completely black except for your photo. Your audience was captive, with no choice but to listen to every word you had to say about your photographic art.

Slideshows are a lost art. Literally. We'll never again have an audience's undivided attention.

That's the most important lesson to teach in this chapter. Slideshows still have their uses, but today, people expect to click, or better yet scroll, to see more pictures. They need to control the pace themselves, or they flip to the next web page or reach for their smartphones. They expect to be able to rapidly scan dozens of pictures instantly and explore only those pictures that catch their attention.

You can use the Slideshow module to make a video of your images, but I don't recommend using slideshows for online presentations because viewers can't control the pace. The Slideshow module is still useful for showing a planned set of pictures to a friend or client, or for filling a screen at a party, but the features are rather limited. For example, it lacks the ability to pan and zoom through pictures.

Though it's not designed for the purpose, you can use the Slideshow module to create a time lapse video, discussed near the end of this chapter.

THIS MIGHT BE THE ONLY SLIDESHOW TIP YOU NEED

For a quick slideshow, stay in the Library module and follow these four steps:

1. Select your pictures.
2. Press **F**.
3. Optionally, fill the screen with your picture by pressing **Ctrl-+** (on a PC) or **Cmd-+** (on a Mac).
4. Scroll through the pictures with the cursor keys.

You can show or hide the metadata by pressing the **I** key. This tip is good enough for most practical uses of the slideshow, and you might never have to read the rest of the chapter.

THE IMPROMPTU SLIDESHOW (THAT I DON'T RECOMMEND)

Sometimes, you just want to show off a few pictures to someone standing over your shoulder. Lightroom has a tool for that purpose: the impromptu slideshow. I suggest you never use it, but just in case you want to try it out, I'll tell you how. At the end of this section, I'll show you a better way to do a slideshow.

The Toolbar on both the Library and Develop modules has a Slideshow button that's hidden by default. Click that arrow on the right side of the Toolbar to select **Slideshow**. If you don't see the Toolbar, press **T**.

Now, click the Play button on the toolbar to start an impromptu slideshow.

Prepare to be frustrated! On my powerful desktop computer, Lightroom took a full minute to prepare a 20-picture slideshow. The slideshow ignores the images you've selected, so you need to create a separate collection just for your "impromptu" slideshow. Once the slideshow does start, there's no way to control it, and if you accidentally click your mouse, the slideshow ends. It's awful, and you're better off leaving the Slideshow button hidden.

Instead, remember the tip from the introduction: Select your pictures and press **F**.

CREATING YOUR SIDESHOW

The Slideshow module ignores selected pictures; your slideshow will always consist of all pictures currently visible in the Library module. Additionally, you might want to flag, crop, or add titles to your pictures while making your slideshow. To be sure you don't edit your original pictures, I suggest creating a new collection with virtual copies of the pictures you want in the slideshow, as shown next.

Next, use the Template Browser in the upper-left corner to find a template that you like. Use the other panels (described later in this chapter) to tweak the appearance.

Use the arrows on the Toolbar to browse through your slideshow and make any modifications to individual pictures. Because you created virtual copies, you can switch to the Develop module to change the crop without impacting your original pictures. If you don't see the toolbar, press **T**.

If you're using a template that shows the title or caption (as shown next), make sure that every picture has a caption filled in. Use the Library module's Metadata panel to add a caption. If you don't want to show the caption or the star rating, simply click the element in the slideshow and delete it. Deleting it for one picture will delete it for all pictures, however.

Use the Playback panel to adjust the pace of your slideshow and to pick which monitor (if you have multiple screens) the slideshow plays on. If you want to control your slideshow, select the Manual Slideshow checkbox, shown later in this chapter.

You're ready to review your slideshow. In the lower-right corner, click **Play** (shown next) and then wait while Lightroom prepares your slideshow.

If you want to make a video, click **Export Video** in the lower-left corner and then specify a filename. Exporting a video can take quite a long time.

The sections that follow provide detailed information about settings available in each panel.

CHANGING THE ORDER OF YOUR PICTURES

To change the order of your pictures, select the collection containing your pictures and then switch to the Library Grid view (keyboard shortcut: **G**). Then, simply drag-and-drop your pictures into the correct order. Unfortunately, you can't reorder pictures in the Slideshow module; you must switch to the Library module.

THE OPTIONS PANEL

The vaguely named Options panel (shown next) controls three aspects of the slideshow's appearance.

Your options are:

- **Zoom to Fill Frame**. Select this checkbox to have Lightroom automatically crop all your pictures to fit the frame, rather than leaving extra blank space around your images. If you do select this checkbox, be sure to scroll through every picture to verify the cropping isn't uncomfortable. You must use the same setting for all pictures; if that doesn't work for your slideshow, manually crop your pictures to 16:9 (or whatever the aspect ratio of the slideshow is) and clear the checkbox. Use your mouse to drag a cropped picture inside the frame for a more pleasing composition.

- **Stroke Border**. This outlines your picture. Click the color box to choose a different color. If you choose a thick Stroke Border, it can look like matting.

- **Cast Shadow**. With this selected, Lightroom shows a shadow behind your picture. Use the slides to control the angle and size of the shadow.

THE LAYOUT PANEL

Use the Layout panel (shown next) to control the size of the borders and the aspect ratio of your pictures.

By default, the **Link All** box is selected, which forces you to have the same sized border on all sides. If you want a different aspect ratio (for example, square pictures instead of 16:9 pictures), clear **Link All**, and then change the **Left** and **Right** or **Top** and **Bottom** sliders as needed.

THE OVERLAYS PANEL

Use the Overlays panel to add an identity plate, watermark, star rating, and text to your slideshow. The next example is deliberately overdone to show the identity plate, watermark, and caption over a picture. I couldn't bring myself to display rating stars even for the sake of the example.

For detailed information about creating an identity plate, refer to Chapter 13. For information about using watermarks, refer to Chapter 15.

Use the **Text Overlays** checkbox to enable or disable all text on your slides. You can separately choose the font style and use the opacity slider to control whether your text is partially see-through.

You can't use the Text Overlays section of the panel to change the font size. Instead, select a text box and drag the corners to resize the box, as shown next.

THE BACKDROP PANEL

Use the Backdrop panel to control the space around your picture. The backdrop is only visible if you have the borders greater than zero in the Layout panel.

Lightroom provides three layers of backdrop, each of which can be turned on or off: **Color Wash**, **Background Image**, and **Background Color**.

Background Color fills the background with a single, solid color. For the next example, I chose an awful red color to make the background more obvious.

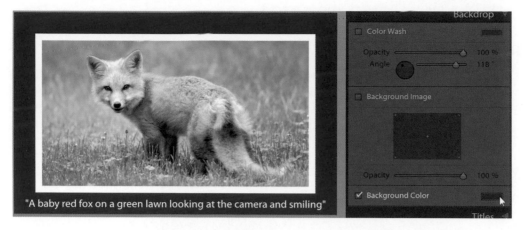

"A baby red fox on a green lawn looking at the camera and smiling"

Color Wash paints a gradient that starts strong at one edge of the picture and fades gradually as it approaches the far corner. The next example adds a blue color wash that starts in the upper-left corner. As you can see, the upper-left corner is pure blue, while the lower-right corner shows the pure red of the background color.

"A baby red fox on a green lawn looking at the camera and smiling"

The **Background Image** fits below the color wash but above the background color. If you have the color wash **Opacity** set to 100%, you'll only see the background image in the more faded parts of the wash. If you use a background image and set the opacity to 100% (shown next), the background color will not show at all.

"A baby red fox on a green lawn looking at the camera and smiling"

For best results with a background image, use a simple image filled only with texture, such as a photo of the sky or grass. I used an abstract picture Chelsea took using the techniques shown at *sdp.io/abstract*.

THE TITLES PANEL

The Titles panel adds a slide before and after your slideshow. You can change the background color and the size of your logo, and that's it. For detailed information about creating an identity plate, refer to Chapter 13.

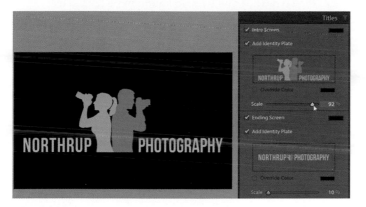

THE PLAYBACK PANEL

Use the Playback panel (shown next) to add music, choose a screen to display the slideshow on, and choose whether the slideshow is timed or changes when you press a key, whether the order is set or random, and whether the slideshow repeats.

When adding music, click **Fit To Music** to evenly space the images across the entire song. If you're including a video in your slideshow, the **Audio Balance** slider controls which audio is louder. You'll need to listen to the playback and adjust the slider to your own taste.

If you have multiple monitors, click a monitor to change which monitor the slideshow uses. **Blank Other Screens** turns all other monitors black during a slideshow; clear this checkbox to show other windows.

ADDING TEXT

Use the **ABC** link on the toolbar to add text to your slideshow. If the toolbar isn't visible, press **T**.

Clicking the link opens a list of standard text options, shown next. If you want to simply type text, select **Custom Settings** and type the text in the text box on the toolbar.

If you select Edit, Lightroom opens the Text Template Editor. This works similar to the Filename Template Editor, discussed in Chapter 15. In a nutshell, select a type of data and then click **Insert** to add it to the text template, or type custom text that will appear on every slide.

Text must be the same on every slide of your slideshow. If you want to show different text for each picture, use the image metadata. For example, type a different Title or Caption for each picture, and then display that using the text tool.

SAVING YOUR SLIDESHOW

Preserve every aspect of your slideshow, including the photos, by clicking **Create Saved Slideshow** in the upper-right corner (shown next). This creates a new collection which is visible from within any module.

You can also save your settings for use in future templates by clicking the + symbol on the Template Browser panel. Then, simply name your template.

MAKING A VIDEO

You can view your slideshow by clicking Play in the lower-right corner. If you want to upload your slideshow, click **Export Video** in the lower-left corner.

Unfortunately, the video options are a bit dated. You can only save at resolutions up to 1080p, even though sites like YouTube allow 4k video with four times the resolution. 60 frames per second (fps) isn't supported, so if you've included video clips in your slideshow, you'll lose the extra quality.

For information about including video clips in a slideshow, refer to Chapter 17.

MAKING A TIME LAPSE

Time lapse videos show a series of timed pictures as a video; showing you traffic, clouds, and stars moving at incredible speed.

You can use the Slideshow module to make a time lapse. First, put your camera on a tripod and configure your camera to take pictures continuously. Exactly how you do this depends on your camera; many newer cameras have an intervalometer built into them. Other cameras require a remote shutter trigger. For more information about taking timed pictures, watch the video at *sdp.io/StarTrails*, the technique is exactly the same for time lapses, and I demonstrate a time lapse in the video.

Once you capture your photos, place them into a collection within Lightroom. If you make any adjustments, be sure to synchronize the changes to all your photos, or the time lapse won't seem smooth.

To create a time lapse, you need to download a custom template, such as the template available at *sdp.io/LRTL*. Extract the template to a file, and then import the template by right-clicking a folder in the Template Browser and selecting **Import**, as shown next.

After importing the template, select it. Then, click **Export Video** and save your time lapse video.

SUMMARY

The Slideshow module is a quick and easy way to present your pictures. If you need more powerful features, consider Photodex Proshow (available at *www.photodex.com/proshow*). Though it's not free, it integrates well with Adobe Lightroom.

By downloading a custom template, you can also use the Slideshow module to create a time lapse. As with the slideshow feature, time lapses work, but it isn't the most robust tool. It's limited to 1080p, and it lacks features common to more powerful time lapse applications that pan, zoom, and remove flickering. LRTimelapse, available at *lrtimelapse.com*, integrates with Lightroom and provides more powerful features, but it also isn't free.

21

Printing Photos & Creating Custom Layouts

To view the videos that accompany this chapter, scan the QR code or follow the link:

SDP.io/LR5Ch21

The Print module is useful in two scenarios:

- Making prints at home with your own printer.
- Creating custom layouts with multiple pictures and text.

This chapter will provide an overview of both scenarios and an overview of soft-proofing. Soft proofing helps you precisely tune your picture for printing, even if you're sending your picture to a printing service.

SHOULD YOU MAKE YOUR OWN PRINTS?

For most amateur and professional photographers, the answer is no. Online photo printing services are a better choice for most of us, because:

- **Online services are cheaper**. It might seem like it would be less expensive to do it yourself, but the cost of paper and ink is usually higher when you print at home.
- **Online services have better quality**. Their commercial printers create pictures that last longer and won't fade as quickly in the sun. They're typically coated, protecting the picture and adding a nice finish. The engineers know how to balance your picture's contrast and color for the best results.
- **Online services are easier**. When I do print my own pictures at home, I rarely get it perfect the first time, so I spend time reprinting. Additionally, I also need to cut my pictures, which takes time and requires a paper cutter.

There are some great reasons to make your own prints:

- **You need them immediately**. Local one-hour services typically provide awful results; you can do better with a good quality printer and photo paper.
- **You need complete control**. If you require absolute control over the brightness and saturation of your picture, it might be easier to do it at home so you can get to know your printer and reprint after making adjustments.

PRINTING SINGLE PICTURES

To make a single print, select your picture and then select the Print module. In the Template browser, select the print size you'd like to make, as shown next. For example, printing a 35mm standard picture at 8x10 requires cropping one inch off the short edges of the picture. Drag your picture to prevent your original composition from being ruined.

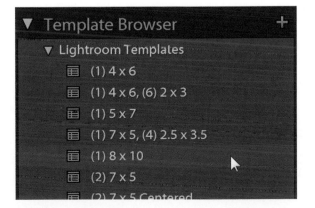

If you just want to print a picture as big as possible, select the **Zoom to Fill** checkbox in the Image Settings panel, shown next.

If the aspect ratio doesn't match that of your picture, grab the picture with your mouse and drag it to control the crop.

CHOOSING YOUR PAPER, SETTING UP YOUR PRINTER, AND PRINTING

Now, you need to setup your printer and paper. The first step is the same for everyone: click the **Page Setup** button in the lower-left corner, as shown next. Then, select the **Paper Size**.

Use the Print Job panel to adjust the print quality, as shown next.

The Print Job panel gives you these options:

- **Print to**. The Print module isn't just for printing; you can use the **Custom Package** layout style for light design work including images and text, and then save the results as a picture.

- **Draft Mode Printing**. I screw up just about every picture the first time I print it. That can be expensive, so I'll often select this checkbox and make a test print. Draft mode printing uses less ink, saving money.

- **Print Resolution**. 300 PPI (Pixels Per Inch) is considered ideal by many and represents the limit of human vision. 200 PPI is usually good enough that most people won't notice a difference.

- **Print Sharpening**. It may seem odd to add even more sharpening than you applied in the Develop module, but print sharpening really does make your pictures look better. I prefer the **Standard** level of sharpening for most photos, or the **High** level for wildlife photos. Be sure to set the **Media Type** to **Glossy** or **Matte**, depending on your paper type.

- **Color Management**. Color Management should usually be set to **Managed by Printer**; this communicates with your printer to create images as close as possible to what you see on the screen. For more information, refer to "Soft Proofing" later in this chapter.

- **Print Adjustment**. If you print a copy and it looks dark or flat, adjust the **Brightness** and **Contrast** sliders and reprint it. If you need to make more complex changes, create a virtual copy and edit it using the Develop module.

If you're using photo paper (which you should for best results) you might need to edit your printer's properties to choose the best settings. You can access these properties by clicking the **Printer** button in the lower-right corner of the Print module. These settings are different for every type of printer and configure exactly how the printer puts ink onto your paper.

In most cases, Lightroom will automatically set the margins correctly for your printer. Most printers can't print to the edge of the paper; thus, these margins are required. You can manually adjust them using the Layout panel, as shown in the next example.

Finally, select the **Print** button (at the bottom-right). Avoid disturbing your printer while it's printing; moving the printer could cause a streak in your picture. After your picture prints, allow it to sit for several minutes until the ink is dry; otherwise, you might smear the picture.

PRINTING CUSTOM-SIZED PICTURES

If you need to print a non-standard size, adjust the **Cell Size** in the Layout panel, as shown next. If you want to print multiple copies, increase the **Page Grid** values.

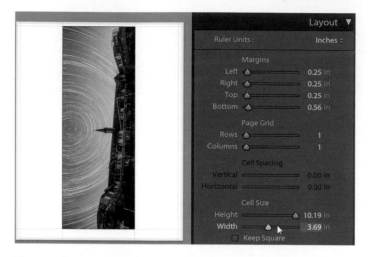

To save these settings for later use, click the + symbol on the Template Browser panel and give your new template a name.

PRINTING MULTIPLE PICTURES ON A PAGE

If you need to print multiple, standard-sized copies of a picture, use one of the existing Lightroom templates, such as **(2) 7 x 5** or **4x5 Contact Sheet**.

If multiple pictures can fit, you can save paper by printing multiple copies of a picture on a single page. In the Layout panel, increase the **Page Grid** values to specify how many pictures will fit on a page. The next example shows how to print two 4x6 pictures by configuring the Layout panel to have 2 rows and one column.

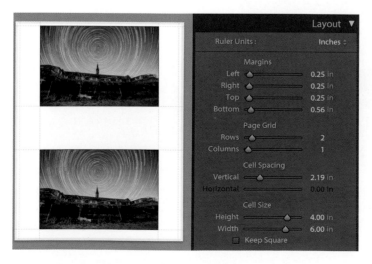

If you're printing the same picture multiple times, select the Repeat One Photo Per Page checkbox on the Image Settings panel, shown next. If you clear that checkbox, Lightroom will fit all selected pictures into the printing template.

PRINTING PICTURE PACKAGES

Picture packages print multiple pictures of different sizes on a single page. Lightroom includes templates for a couple of picture packages, including **(1) 4 x 6, (6) 2x3**, shown next.

Note that picture packages always print the same picture in every available spot on a single page. If you select multiple pictures, Lightroom will print them on separate pages. If you want to mix different pictures on a single page, create a Custom Package instead. I describe custom packages later in this chapter.

Use the Cells panel to create custom picture packages. First, click **Clear Layout**. Then, click the different button sizes to add pictures to the page. To add a different picture size, click the triangle next to an existing size, click **Edit**, and add your own size.

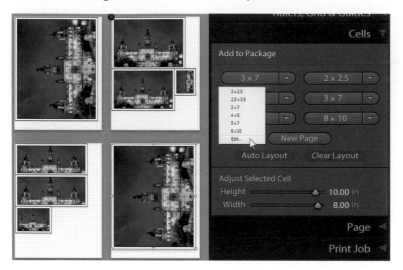

Lightroom will automatically arrange the picture sizes you select to fit on the page in the most efficient way possible. You can also drag the pictures with your mouse.

If you want to use the same picture package in the future, remember to save it as a template.

WATERMARKING YOUR PRINTS

Though many professional photographers today provide digital proofs, some photographers still print contact sheets or other samples for clients. If you're printing a sample and you still hope to sell your your client other prints, you might want to add a watermark.

The Page panel in the Print module allows you to select a watermark, along with a few other options. To select a watermark, select the **Identity Plate** checkbox. Then, click the number to the right (as the next example shows) to rotate your identity plate. Select the **Render on every image** checkbox. Adjust the **Opacity** and **Scale** as needed. For more information about creating an identity plate, refer to Chapter 13.

The Page panel also has a **Watermarking** option. The two are very similar. The key difference is that Lightroom centers the Identity Plate and allows you to rotate it. Watermarking is also flush against the bottom of the picture. For more information about watermarking, refer to Chapter 15.

MAKING CUSTOM LAYOUTS

You can also create your own print designs, placing pictures anywhere on the page, overlapping images, and adding text. Custom packages aren't intended for printing individual pictures that will be cut and framed; the feature is intended for turning the entire page into a work of art.

CUSTOM PACKAGES

The next example shows the **Custom Overlap x3 Landscape** template. After selecting a custom template, drag your pictures from the filmstrip at the bottom of the page into the boxes. To adjust the crop of a picture, hold the **Ctrl** key (on a PC) or the **Cmd** key (on a Mac) while dragging the photo.

You have complete control over the layout. For example, if I want to move a picture to the front, I can right-click it and then select **Send To Front**, as shown next.

Your options for layout are a bit limited; there are no options for drop shadows or rounded edges. You can use the Image Settings panel to add borders to your image, however. The **Photo Border** option adds a white border, while the **Inner Stroke** option allows you to select a different color.

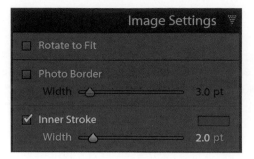

If you'd like to put your logo anywhere on the page (and not be restricted to the Identity Plate or Watermarking options), import your logo into Lightroom as a photo and add it to the Custom Package. Selecting the option for borderless printing in your printer settings can allow you to move pictures closer to the edges.

To save the set as its own picture, use the **Print Job** panel with **Print to** set to **JPEG File**. The Print Job panel gives you vague control over the size of the exported picture by using the **File Resolution** setting. To determine the PPI that you need, divide the desired pixels by the width of the page in inches. For example, if the paper is 8.5x11" and you want the final picture to be 1000 pixels wide, divide 1000 by 8.5. Set the **File Resolution** to the result, which in this case is 118. Finally, click **Print To File** in the lower-right corner.

PLACING YOUR LOGO ANYWHERE

Lightroom is not very flexible about where you can place your logo; either you use the watermark feature and place it over every picture, or you use an identity plate and drop it right in the middle of the picture.

You can overcome this by adding your logo to a custom package and positioning it off-center. Follow these steps:

1. Import your logo into Lightroom, if necessary.
2. In the **Layout Style** panel, select **Custom Package**.
3. In the **Cells** panel, click **Clear Layout**.
4. Click **Page Setup** and select the page size that you plan to use for your final layout. You can repeat this process for different page sizes.
5. Drag your logo to the page and position it as you want it to be positioned in the final picture, as shown next. You're just adding your logo to the page in this step, so leave room for your pictures elsewhere on the page.

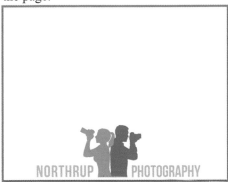

6. In the **Print Job** panel, click the **Print to** list and select **JPEG File**.
7. Click **Print to File** and save your picture.
8. Switch to the Library page and drag your newly saved, formatted logo into the Lightroom library.
9. Now, create a new print layout. In the next example, I used these settings:
   - **Layout Style panel**: I selected **Single Image/Contact Sheet**.
   - **Page panel**: I selected **Identity Plate**, clicked **Edit**, and chose the newly made offset logo. I also selected the **Render behind image** checkbox.
   - **Layout panel**: I adjusted the margins to add space below the pictures for the logo, as shown next.

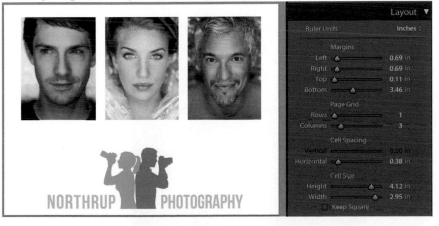

Now, your custom layout is ready for printing!

CREATING DESIGNS WITH YOUR LOGO

Now that you have your logo saved as a formatted image, you can use it as an identity plate without it being centered.

SAVING PRINTS

You can save a layout for later by clicking the Create Saved Print button at the top of the page. This creates a collection that includes any images used in the print job.

SOFT PROOFING

If you have a high-quality printer and you do your own printing, you've probably experienced making a print and having it look very different from what you saw on screen. Then, you made some adjustments in Lightroom and reprinted the picture. Good news: soft proofing can save you some ink

Soft proofing gives you a better idea of what a physical print will look like before you print it. It's only useful to photographers with calibrated monitors and printer profiles; if you send your picture to a service for printing (which is what I recommend to most people), the printer will take care of the adjustments. That is to say, most of us will never need to use soft proofing.

To use soft proofing, select the Develop module. On the toolbar (press **T** to display it if it's hidden), click the triangle and select **Soft Proofing**, if necessary. Then, select the **Soft Proofing** checkbox.

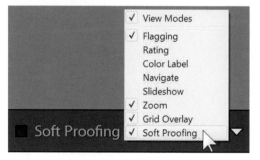

Now, the Soft Proofing panel replaces the Histogram panel in the upper-right corner of the window, as shown next. Lightroom also changes the background of your photo to white (the most common color for a print matte). To change the color, right-click the background and pick a different shade of grey, as shown next.

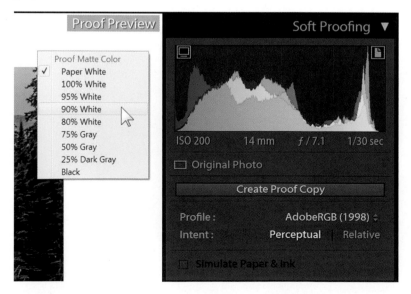

In the Histogram, notice that the upper-left and upper-right corners have changed. Click the monitor icon in the upper-left corner to show parts of the picture that your monitor is incapable of displaying. Click the paper icon in the upper-right corner to highlight parts of the print that the selected printer or color space won't be able to properly render.

Usually, soft proofing reveals that parts of your picture are too dark or saturated for your printer. You don't necessarily need to change these parts of the picture; your printer will do its best to render the picture anyway. However, if a large part of your picture is as saturated as your printer can possibly render, you might not be able to see texture and detail in the saturated areas; it might appear solid. The next example shows that the printer selected in the Profile list can't render detail in the brightest highlights of the flower, which Lightroom has highlighted in red.

If you're concerned about losing detail in highly saturated parts of your picture, decrease the saturation using the adjustment brush, as described in Chapter 6. If you're concerned about losing detail in shadows, raise the shadows.

Soft Proofing has a handful of other options:

- **Profile**. It's important to select the printer that you'll be using. Look for a .ICM or .ICC profile for the printer, install it on your computer, select **Other** from the **Profile** list to enable that profile, and then select the printer from the list.

- **Intent**. Perceptual is usually the right choice; it attempts to render the colors so that the human eye will see them naturally. Relative bases the display on the source color space's extreme highlights. You might see more original colors using relative than perceptual, but perceptual is usually more useful.

- **Simulate Paper & Ink**. Whites are never pure white in a print, and blacks are more of a dark grey than a black. Selecting this checkbox (which is only available when a printer profile is selected) will simulate this.

SUMMARY

The Print module is good for more than just basic printing; you can use it to create custom print packages and gorgeous, modern designs. For more complex layouts, or special effects such as drop shadows, you might consider exporting your pictures into Photoshop.

22
The Web Module

To view the videos that accompany this chapter, scan the QR code or follow the link:

SDP.io/LR5Ch22

The Web module promises a quick way to publish an album of photos to the Internet. There was a time when the Web module was efficient, beautiful, and even exciting. That time passed many years ago, however, and Adobe never bothered to update the Web module.

- I can't think of a single scenario where I would recommend using the Web module. Here's why:
- The templates are non-scalable, so your pictures are far too small on many new displays.
- The sites are not mobile-friendly, so they'll look awful and work terribly on mobile devices.
- The designs are outdated, which will make even a new photo album look old.
- There are better, faster, and free ways to show off your pictures online.

Instead of the Web module, I recommend using one of many different websites available on the Internet. These websites offer publishing plugins that make it simple to share your photos:

- Flickr.com
- 500px.com
- Facebook.com

And one final tip for this chapter: right-click Web in the top-right corner of the Lightroom window and clear the Web checkbox to hide the module forever.

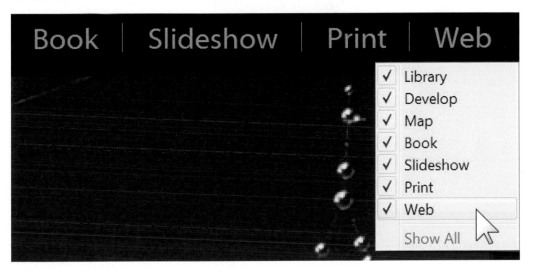

If you really must learn about the Web module, or you just want to see an angry nerd complain for fifteen minutes, watch the Web module video at *sdp.io/LR5Ch22*.

23
Backups, Catalogs, & Previews

To view the videos that accompany this chapter, scan the QR code or follow the link:

SDP.io/LR5Ch23

This chapter isn't as exciting as applying fake makeup to a model, but it might be the most important chapter in the entire book. This chapter can save you from genuine heartache by teaching you to manage your Lightroom catalogs and keep your pictures backed up. I'll also show you how to manage previews, which can drastically improve Lightroom performance.

BACKING UP YOUR PICTURES

First, backups are very important. The hard drive in your computer will fail at some point, and if you don't have a backup of your pictures and your Lightroom catalog, you will lose all your pictures.

Your hard drive *will* fail. Your computer might also be stolen, or damaged in a fire, or hacked across the Internet, or soaked by a flood. I've gotten emails from dozens of absolutely heartbroken photographers asking me how they can recover files from their failed hard drive—all because they never bothered to setup a backup.

I'm being redundant because this is important: you must have your pictures in at least two places or you will lose them all.

OFF-SITE BACKUPS

Backing up to an external hard drive is good, and it protects you from drive failure, but it doesn't protect you from theft, fire, or natural disaster. For most people, the easiest way to back up your pictures and protect them from most scenarios is to use an online backup service. Some of the most popular services are (in no particular order): Carbonite.com, Livedrive.com, Backblaze.com, SOSonlinebackup.com, CrashPlan.com, Mozy.com.

I'm not endorsing any of these services. They change prices and features too frequently for me to provide a recommendation.

LOCAL BACKUPS

Not everyone understands this, so I have to say it explicitly: you shouldn't back up your files to the same hard drive that you store them on. If that hard drive fails, it will probably fail completely, preventing you from accessing either the data or the backup.

Instead, you should back up your pictures to a second hard drive. The easiest way to install a second hard drive is to buy an external hard drive that connects using a USB connection. For best results, get a USB 3 hard drive, and for backups, don't bother paying extra for a Solid-State Drive (SSD).

When shopping for external hard drives, you'll see two different types:

- **Desktop external hard drives**. Desktop hard drives are bigger, higher capacity, and less expensive. They're better for backups when you can reach a power outlet and you don't mind the separate power plug.
- **Mobile hard drives**. Mobile hard drives tend to be smaller, lower capacity, and slightly more expensive, and they don't require a separate power supply. Use a mobile hard drive when you travel because they're easier to carry.

You can use a mobile hard drive for your day-to-day backups, but if you do that, you shouldn't take it with you when you travel. If you travel with both your computer and your backup drive, you'll lose everything if you forget your bag in a taxi or someone robs your hotel room. Leave a backup drive at home if you don't have off-site backups.

MANAGING PREVIEWS

If you wish browsing pictures in Lightroom were faster, you need to learn about previews. Previews take time to generate, but they make Lightroom much quicker.

When you select a picture in Lightroom, the software could read your raw file from the disk, apply all your adjustments to it, and then show it to you. That would take about as long as it took to originally import your picture, and that's long enough to be really annoying. In fact, if you don't have any previews, that's what Lightroom does.

Previews make Lightroom faster by storing a smaller, temporary version of your picture just for the purpose of speeding up browsing. Following in Adobe's long tradition, managing these previews is unnecessarily complex, and if you fail to manage them efficiently, you can fill up your disk and cause everything to break.

The sections that follow describe Lightroom's three types of previews, and yes, you really do have to understand the difference.

STANDARD PREVIEW

Standard Previews are small copies of your pictures that make it faster to browse your pictures when you aren't zooming all the way in. You see them when you use the Fit option to display a full picture in the Loupe view. If you don't generate standard previews when importing pictures, Lightroom generates them automatically when you display a picture.

Standard previews aren't as big as the original picture, so Lightroom can't use them when you zoom in to a picture. They're useful to speed up browsing pictures, though.

You can control the size of standard previews. You should select a size that's at least as big as your monitor is wide, in pixels. For example, if you have a widescreen 1920x1080 monitor (a common resolution) you might want to select a standard preview size of 1440 pixels. To change the standard preview size, open the **Library | Catalog Settings** (on a PC) or the **Lightroom | Catalog Settings** (on a Mac) menu and select the **File Handling** tab, as shown next.

1:1 PREVIEWS

These previews allow you to zoom in 1:1 to a picture. If you know you'll need to zoom in when culling your pictures (for example, to find the sharpest wildlife pictures from your set), you should generate 1:1 previews to make the process faster. 1:1 previews can take up massive amounts of your disk space, but Lightroom can automatically delete them, so it only stores 1:1 previews for your recent pictures (which are usually the ones you access most).

SMART PREVIEWS

These previews allow you to view and edit your pictures even if your pictures are stored on an external hard drive that isn't currently connected. If you have a laptop and store your pictures on an external drive, you can use smart previews to view and edit lower-resolution pictures even when your external drive isn't plugged in.

Basically, it makes a smaller copy of your picture—about 95% smaller than your original raw file—but with reduced quality. The smart previews are stored with your catalog, not your pictures. If you're working with a laptop, be sure to store the catalog on your local disk, even if you store pictures on an external disk.

The "smart" aspect of these previews is that Lightroom automatically uses the previews when your original pictures aren't available. As soon as you reconnect your external drive, Lightroom will use your original files and ignore the smart previews.

You can export pictures based on the smart previews when your external drive isn't connected, but they'll be lower resolution: usually 2,048 pixels wide.

You can use the Histogram to see if a picture has a Smart Preview available. Click the message to delete the Smart Preview, as shown in the following figure.

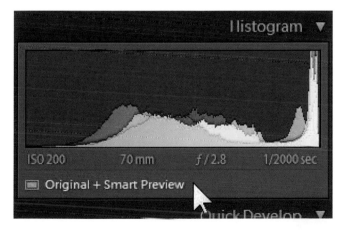

ADVANCED: USE SMART PREVIEWS TO SPEED UP EDITING

If you shoot raw, especially with with a high-resolution sensor (like the 36 megapixel Nikon D810 or Sony A7R), Lightroom can be annoyingly slow, even with a fast computer.

You can use Smart Previews to speed up the import and editing process:

1. When you import your raw files, configure Lightroom to automatically generate smart previews. You don't need 1:1 previews, which will also speed up editing. Make note of the folder Lightroom imported the pictures into.

2. Close Lightroom.

3. Use Explorer (on a PC) or Finder (on a Mac) to rename the folder. For example, you might add "-temp" to the current folder name.

4. Re-open Lightroom and do your processing and editing. You'll notice that it's much faster because Lightroom can't find the original raw files and is instead working with the smaller smart previews.

5. When you're ready to export your pictures, close Lightroom. Then, rename the folder to its original name and restart Lightroom. Now, Lightroom will locate your original raw files and apply all your edits to the originals.

Thanks to Horshack on the DPReview forums for discovering this trick!

GENERATING PREVIEWS DURING IMPORT

The easiest way to generate previews is when you import pictures, as shown in this figure.

You have four options:

- **Minimal**. The Minimal option imports your pictures as fast as possible by not rendering a standard preview. With this option selected, Lightroom waits until you view a picture in Loupe to render the Standard preview. Minimal is the best choice if you want to copy pictures to your computer but you don't plan to ever look at most of them.

- **Embedded & Sidecar**. This option uses a preview that might or might not be built into image from the camera. You don't ever need to select this.

- **Standard**. Select Standard if you plan to browse through your pictures but don't plan to zoom in to 1:1 to examine the sharpness.

- **1:1**. If you do plan to examine the details of the pictures, select 1:1, and then grab a cup of coffee after you click the Import button—because Lightroom is going to be processing pictures for quite a while.

Notice that Smart Previews aren't shown on that list. There's a separate checkbox to enable Smart Previews that you should select only if you're importing the pictures to an external drive that you might later disconnect.

GENERATING PREVIEWS AFTER IMPORT

If you didn't generate previews when you imported your pictures, or you want to generate 1:1 previews for specific images, select the images and then choose an option from the **Library | Previews** menu.

DISCARDING PREVIEWS

You can discard previews in the same way; select them and choose an option from the **Library | Previews** menu. You can also configure Lightroom to automatically discard old 1:1 previews by opening the **Edit | Catalog Settings** menu item (on a PC) or the **Lightroom | Catalog Settings** menu item (on a Mac) and selecting the **File Handling** tab, as shown in the Standard Preview section earlier in this chapter.

USING CATALOGS

A Lightroom catalog is a physical file that stores your edits and metadata. It doesn't actually store your photos, just the changes that you've made. Catalogs have a .lrcat file extension.

By default, Lightroom creates a catalog at Pictures\Lightroom\Lightroom 5 Catalog.lrcat. Most home users will never need to create another catalog, and thus you can simply skip to the next section.

Professional users might need to create multiple catalogs, however. Creating a new catalog is a great way to keep pictures completely separate, preventing you from accidentally mixing pictures. It's also the easiest way to pass pictures between different Lightroom users. For example, after a shoot, I create a new catalog for my favorite photo editor, Chelsea. I can store the catalog on an external drive (along with the pictures), and simply hand the drive to her. She can connect it to her computer, see all my ratings and edits, and do her own editing. Then, she can hand it back to me, and I can pick up where she left off.

Even if you're a professional user, you should probably avoid creating multiple catalogs. Managing all your pictures in a single catalog is difficult, but it becomes even harder when you divide it into multiple catalogs. Pictures in different catalogs aren't searchable in any way, so to find a picture, you have to remember where you stored that other catalog.

It's nice that Adobe allows us to create multiple catalogs, but it also needs to provide tools to allow us to organize and manage multiple catalogs. Until that becomes a feature, I recommend most users put all their pictures into a single catalog.

In the sections that follow, I'll show you how to create, export, and open catalogs.

CREATING A NEW, EMPTY CATALOG

If you want to create a new catalog to import pictures into, open the **File** menu and then select **New Catalog**. The new catalog will be empty, so you'll want to import some pictures, as discussed in Chapter 1.

EXPORTING A NEW CATALOG

If you have pictures in an existing catalog that you want to separate into their own catalog, select the pictures, open the **File** menu, and then select **Export As Catalog**. You have four options when you export a catalog:

- **Export Selected Photos Only**. Clear this checkbox if you want to export all the pictures currently visible in the Grid view, but you accidentally had only a few pictures selected.

- **Export Negative Files**. Select this checkbox to export your pictures with the catalog. If you're passing the catalog to someone who doesn't yet have the pictures, leave this selected. If they already have the pictures and you just need to export your metadata and Lightroom edits, you can clear this checkbox to save space.

- **Build/Include Smart Previews**. Smart previews allow you to view and edit pictures even if the original file isn't available. If you're passing someone a catalog, they don't have the original pictures, and the original pictures are too big to include with the catalog, select this checkbox. The smart previews are smaller than the original pictures, and allow someone to edit the pictures even if they don't have them, but they're not suitable for exporting or making prints.

- **Include Available Previews**. If you've previously rendered previews other than smart previews (such as 1:1 previews), select this to include them. That will increase the size of the catalog, but make it faster to work with the pictures when you open the newly created catalog. If you don't include available previews, you can create the previews after opening the new catalog.

Don't worry too much about these checkboxes; if you just leave the defaults selected, you're probably fine.

OPENING A CATALOG

You can open a catalog from Explorer (on a PC) or Finder (on a Mac), just like any other file. You can also use the File menu and browse the Open Recent list, or just click Open Catalog to browse your file system.

By default, Lightroom opens your most recently used catalog when you start it.

BACKING UP CATALOGS

If you use the backup software included with Windows or Mac OS, you'll automatically back up your Lightroom catalog. You don't actually need to use Lightroom's built-in backup feature to be prepared for data loss.

The built-in backup feature is useful for recovering from a potentially corrupted catalog. Backups are useful for that purpose, too, but if you don't use Lightroom every day, you might overwrite your good backup with a backup of the corrupted file. The backup systems included with newer versions of Mac OS and Windows also protect you from this by keeping multiple versions of a file when you have sufficient space on your backup drive. Additionally, I haven't had a corrupted catalog that Lightroom couldn't automatically repair in recent versions of Lightroom, however.

By default, Lightroom makes a second copy of your catalog every week. You've probably seen that annoying dialog that pops up when you try to close Lightroom.

As I mentioned, if you are backing up your computer using software that keeps multiple versions of your files, you can safely turn off that annoying Lightroom backup. From the menu, choose **Edit | Catalog Settings** (on a PC) or **Lightroom | Catalog Settings** (on a Mac), and then pick your backup frequency.

Catalog Settings

General | File Handling | Metadata

Information

Location: C:\Users\tnorthru\Pictures\Lightroom   [Show]

File Name: Lightroom 5 Catalog.lrcat

Created: 6/10/2013

Last Backup: 5/21/2011 @ 4:43 PM

Last Optimized: 9/25/2014 @ 10:01 AM

Size: 1.58 GB

Backup

Back up catalog: | Once a week, when exiting Lightroom |▾|

	Never
	Once a month, when exiting Lightroom
✓	Once a week, when exiting Lightroom
	Once a day, when exiting Lightroom
	Every time Lightroom exits
	When Lightroom next exits

OPTIMIZING CATALOGS

Catalogs require some regular maintenance, or Lightroom will slow down and the catalog will consume unnecessary disk space. Fortunately, Lightroom does this for you automatically, so you should never have to worry about it. However, I'll tell you where the menu item is because some readers like to read about every single feature, even if they never plan to use it.

To optimize a catalog, open the **File** menu and then select **Optimize Catalog**. Then, make a cup of tea, because for some reason Lightroom can't optimize the catalog in the background while you continue to work.

Confirm

Lr   Your catalog was last optimized 9/24/2014 @ 4:43 PM. If it has been running slowly and you haven't optimized it recently, optimizing it again may improve performance.

Optimization may take several minutes. You will be unable to use Lightroom during this time.

[Optimize]   [Cancel]

SUMMARY

Here's what I hope you learned from this chapter:

- Have an off-site backup of your pictures.

- Use previews to make your editing faster, but get rid of the old previews when you run low on disk space.

- Add collections when you need to separate your pictures or hand pictures to an editor.

24
Tips & Tricks

To view the videos that accompany this chapter, scan the QR code or follow the link:

SDP.io/LR5Ch24

Just in case you don't read the entire book, here's a collection of random tips that are easy to overlook but can make a big difference in your photo editing.

DRAG NUMBERS

You can click-and-drag almost any number to make adjustments with your mouse.

DIRECTLY ADJUST PARTS OF YOUR PICTURES

The Tone Curve, HSL, and B&W panels feature the direct adjustment tool in the upper-left corner. Click it, and then click-and-drag part of your picture to adjust that specific part of the picture. That's much easier than trying to guess whether part of your picture is considered "dark" or "shadow."

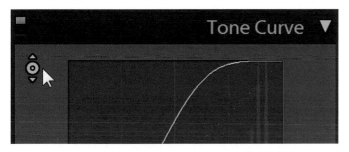

RESET SETTINGS

If you want to see your picture without the changes you've made in a specific panel, click the switch in the upper-left corner of the panel, as shown next. When it's in the down position, all changes made with the panel are hidden. Click it again to reapply those changes.

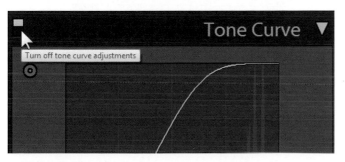

Want to reset changes in almost any panel? Double-click a label. For example, double-clicking Shadows (shown next) instantly resets it to zero. You can double-click almost anything in Lightroom.

You can also hold down the **Alt** key (on a PC) or the **Opt** key (on a Mac), and some labels will change to allow you to reset settings. The next example shows the Basic panel without the **Alt/Opt** key (on the left) and with the **Alt/Opt** key (on the right):

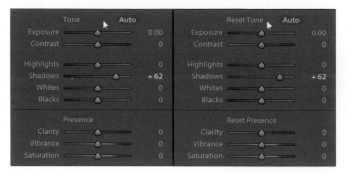

APPLY SHARPENING TO ONLY THE DETAILED PARTS OF YOUR PICTURE

When you apply sharpening to a picture, it will accentuate noise in the blurry background, making it worse. You can use the Mask tool to apply sharpening only to the detailed parts of the picture.

In the Develop module, open the Detail pane. Hold down the **Alt/Opt** key, and drag the **Masking** slider to the right. Your picture will appear black-and-white. Slide the **Masking** slider until the background is mostly black, and your detailed subject is mostly white. Now, you can adjust the **Sharpening** slider up further because Lightroom will only sharpen the detailed parts of your picture that were white in the preview.

MAXIMIZE SCREEN SPACE

When you need more room to see your pictures, you can do several things:

- Press the **F** key to show the picture full screen. This is a quick way to zoom in on a picture when you're browsing the grid view.

- Press **Shift+F** repeatedly to remove the title bar and windows. This leaves panels in place, allowing you to continue editing, but minimizes wasted space.

- Press the **Tab** key to hide all panels. Press it again to show them.

When working with a small screen, I always press **Shift+F** twice to hide the menu and title bars. If you do need to access the menus (a rare task), just press **Shift+F** again. Or, if you really want to be fast about navigating the menus, hold down the **Alt/Opt** key and press the first letter of the menu. For example, press **Alt+F** (on a PC) or **Opt+F** (on a Mac) to display the **File** menu (even when the menu bar is hidden) or press **Alt+E** (on a PC) or **Opt+E** (on a Mac) to see the Edit menu.

MATCHING EXPOSURES

If you use autoexposure, your camera might use different exposures for pictures shot in the same conditions. For example, if you take five pictures at a wedding, the camera might use different exposures for shots centered on the groom's dark suit and shots that fill the frame with the bride's white dress.

Lightroom can examine the camera settings used for each picture and adjust the exposures to match. First, select the entire series of pictures. Then, click the picture with the best exposure. In the Develop menu, use the menu to select **Settings** | **Match Total Exposures**. Lightroom will adjust the exposures so the pictures match.

This is especially useful when you'll be displaying the pictures side by side; changes in exposure are very distracting in a sequence.

SEE CLIPPED WHITES AND BLACKS

It's good if small parts of your picture are completely white or black; that means your picture has excellent contrast. However, parts of the picture that are pure black (0% brightness) or pure white (100% brightness) won't have any detail at all. There are two ways to see exactly which parts of your picture are clipped: using the Histogram and using the Whites and Blacks tools.

USING THE HISTOGRAM

Click the triangles at the upper-left and upper-right of the histogram to highlight the under- and over-exposed parts of the picture, respectively. This example shows the pure black parts of this photo in blue:

After raising the exposure and selecting the triangle in the upper-right corner of the histogram, we can see all the 100% pure white parts of the picture highlighted in red:

USING THE SLIDERS WITH THE ALT/OPT KEY

If you hold down the **Alt** key (on a PC) or the **Opt** key (on a Mac) while using the sliders in the Basic panel of the Develop module, Lightroom will hide everything except the parts of the picture that are underexposed or overexposed (depending on which slider you're using).

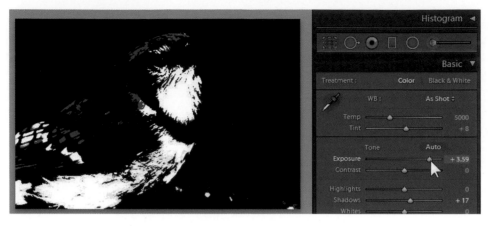

DRAG THE HISTOGRAM

In the Develop module (but not the Library module, for some reason) you can directly change several of the sliders in the Basic panel:

- Drag the far left part of the histogram to adjust the blacks.
- Drag the middle left part of the histogram to adjust the shadows.
- Drag the middle of the histogram to adjust the overall exposure.
- Drag the middle right part of the histogram to adjust the highlights.
- Drag the far right part of the histogram to adjust the whites.

This has the exact same effect as dragging the sliders, but it can be more intuitive. The next figure shows where to drag to adjust the shadows in a picture.

DRAG PHOTOS INTO LIGHTROOM

Have a random photo that you want to edit in Lightroom? Don't bother browsing for it from the Import dialog—just drag it directly into Lightroom. Lightroom will open the Import dialog to the correct folder, with only the selected picture checked.

USE THE SPACE BAR TO DRAG YOUR PHOTOS

If you have a tool (such as the Spot Removal) tool selected but you need to pan/scroll around your picture, just hold the space bar down on your keyboard. This turns your cursor into a hand that you can use to drag your picture.

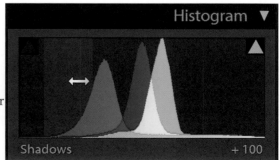

QUICKLY CHANGE THE DEFAULT SETTINGS FOR NEW PICTURES

When Adobe imports raw files, it automatically applies default settings for color, contrast, and exposure. Some people don't like the defaults. Fortunately, there's an easy way to change them.

In the Develop module, edit a picture with subtle adjustments that you want applied to all raw photos from that specific camera model. Then, hold down the **Alt** key (on a PC) or the **Opt** key (on a Mac) and click **Set Default** in the lower-right corner.

Then, click Update To Current Settings. If you have more than one camera, you'll have to repeat this process for every camera model that you have. If you change your mind about this, select a picture taken with the same camera and click Restore Adobe Default Settings.

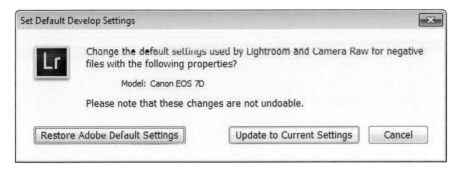

SAVE 15% OF YOUR DISK SPACE WITH DNGS

If you shoot raw, you know they can quickly fill your disk. I recommend converting images to DNG files during import because DNG files are losslessly compressed, which means the files are smaller but there's no disadvantage or lost data. If you've already imported raw files without converting them to DNGs, it's not too late to save the disk space; Lightroom can convert the files to DNGs and delete your original, less-efficient raw file.

This quick process can reduce the size of your photo library by 15%:

1. First, find any non-DNG raw files in your library. Right-click any existing Smart Collection and then click **Import Smart Collection Settings**. If you don't have a single Smart Collection, create a temporary one just so you can right-click it.

2. In the Import Smart Collection Settings dialog, type **http://sdp.io/findraw**. This points to a Smart Collection I made just for you to find all raw files by any camera manufacturer. Click **Open**, and wait a moment.

File name: http://sdp.io/findraw   lrsmcol (*.lrsmcol)

Open   Cancel

3. Click the new Smart Collection. Then, select all the pictures in the collection by pressing **Ctrl+A** (on a PC) or **Cmd+A** (on a Mac).

4. Select the **Library | Convert Photos To DNG** menu item.

5. Configure the Convert dialog as shown next. Select **Delete Originals After Successful Conversion**, **Only Convert Raw Files**, and **Embed Fast Load Data**. Clear the remaining checkboxes. For the JPEG Preview, select **None**. Click **OK**, and Lightroom will begin the potentially long process of converting your inefficient raws.

Convert 10033 Photos to DNG

Source Files
☑ Only convert Raw files
☑ Delete originals after successful conversion

DNG Creation
File Extension: dng
Compatibility: Camera Raw 7.1 and later
JPEG Preview: None
☑ Embed Fast Load Data
☐ Use Lossy Compression
☐ Embed Original Raw File

OK   Cancel

This saved me over 150GB of disk space!

USE RAW FILES FROM YOUR BRAND NEW CAMERA

Every camera has a unique raw file format, and if you get a brand new camera, you probably won't be able to use Lightroom to process the raw files until Adobe releases an update. This is incredibly frustrating, because the first thing any new camera owner wants to do is to take pictures and then spend hours looking at every detail.

It requires an extra step, but there is a way to get those raw files into Lightroom without degrading quality: use the latest version of the Adobe DNG Converter, shown next and available for free at *sdp. io/cameraraw*. Adobe usually releases the DNG converter before a Lightroom update.

If the latest release version of Adobe DNG Converter doesn't support your camera's raw files, check for a newer beta version at *labs.adobe.com*. Adobe's pretty fast about releasing updates after a new camera is publically available.

After converting your camera's raw files to DNG, you can import the DNG files with Lightroom and process them like any other raw file.

GET THAT JPG LOOK

I've convinced many people to switch from shooting JPG to raw. Many of them complain that their raw files don't have the "look" that their JPGs did, though.

Raw is better, no doubt, but your in-camera processing might add a nice tint that makes the colors pop. Of course, you can match any look by adjusting the colors in the Develop module of Lightroom, but there's an easier way, too. In the **Camera Calibration** panel, change the **Profile** from **Adobe Standard** to one of the other options—whichever you like the look of. **Camera Faithful** will be as close as possible to your camera's default JPG settings.

FIND YOUR COLLECTIONS

For some reason, Lightroom doesn't let you search collections by their name. If you're like me and you have hundreds of collections, it might be easier to find a picture within the collection.

Once you find a picture, right-click it, select **Go to Collection**, and then select a collection.

JUMP TO RECENT AND FAVORITE COLLECTIONS

Adobe did a great job of hiding a really useful tool: Favorite Sources. If you click the path name of a collection, Lightroom shows you a menu that you can use to quickly jump back to recent collections. If you select a collection before opening the menu, you can click **Add to Favorites**, as shown next. Favorites always appear in the **Favorite Sources** list near the top of the menu.

1

1:1 Previews, 15, 171, 239, 241-242

A

Adjustment Brush, 63, 66-73, 75, 91, 105-106, 111-112, 233
Apply During Import, 127, 171
Aspect Ratio, 12, 44, 51, 53-54, 56, 98, 129, 216-217, 224
Attribute, 9-10, 36-37, 193
Auto Import, 190-191
Auto Layout, 205, 208-209

B

B&W, 89, 103-105, 107, 109, 111, 141, 245
Backdrop, 217
Background Color, 149, 210-211, 217-219
Background Image, 217-218
Basic Panel, 20, 22, 24, 52, 87, 108-109, 246, 248
Black and white, 44
Blacks, 17, 25, 28, 106, 108-109, 195-196, 233, 247-248
Book Module, 205-207
Book Settings, 206-207

C

Camera Calibration, 24, 96, 124-125, 127, 129, 131, 133-135, 251
Caption, 35-36, 44, 208, 215, 217, 220
Capture Time, 31
Catalog, 4, 47-49, 125, 143, 159-160, 162, 171, 173, 175, 182, 237-239, 241-243
Catalog Settings, 160, 238, 241-242
CD/DVD, 174
Cell, 152, 154-156, 226
Cell Icons, 154
Cells Panel, 228, 231
CF card, 190
Chromatic Aberration, 125, 128-129, 131
Clarity, 8, 17, 28, 51, 64, 71-73
Collections, 40-49, 243, 251
Color Label, 37-38, 154, 156
Color Label Set, 37-38
Color Management, 225
Color Wash, 217-218
Compact Cells, 151-152, 155

Compare View, 7
Copy as DNG, 171

D

Destination, 171-172, 190-191
Develop Settings, 8, 135, 161-162, 171, 189-190, 196
Distortion, 125, 127, 130-133
DNG files, 165, 249-250
Draft Mode, 224

E

Email, 174
Expanded Cells, 151-152, 156
Export, 14-15, 35-36, 142, 148-149, 163, 171, 173-175, 177, 179-181, 183-185, 193, 198, 215, 220-221, 239-242
Export as Catalog, 242
Export Location, 175
Exposure, 7-8, 17, 19, 25, 27, 64-66, 68-70, 73, 75, 106, 109, 111-112, 195-196, 247-248
File Handling, 159, 164, 169, 171, 194, 238, 241

F

File Naming, 165, 175, 190
File Renaming, 171
File Settings, 177
Fill, 24, 69, 108, 166, 198, 207-208, 213, 216, 224, 238, 247, 249
Filmstrip, 30, 38, 150, 166-168, 202, 229
Filter, 9-11, 20, 34, 36, 38, 63-66, 150, 155, 182, 193, 202
Fit, 44, 56, 151-152, 178, 207-208, 216, 219, 226-228, 238
Flag, 6, 39, 44, 153, 167, 214
Folders, 2, 33, 40-41, 43, 45, 47-49, 171, 188
FTP Upload, 174

G

GPS, 44, 178, 201-203
Graduated Filter, 63-65
Grain, 100-101, 115
Grid, 4-5, 7-9, 14, 30, 37, 39, 54-55, 103, 150-151, 153-155, 184, 198, 202, 216, 226, 242, 246
Guides, 54, 56

H

Highlights, 17-18, 24-28, 64, 70, 77-80, 82-83, 85-87, 95-96, 99, 116, 183, 233, 248
Histogram, 17-19, 21, 26-28, 31, 77, 81, 106, 108, 113, 195, 232-233, 239, 247-248
History, 100, 103, 136-139, 141, 143
HSL, 89, 91, 245
Hue, 89, 93-94, 132

I

Identity Plate, 145-146, 217, 219, 228-232
Image Settings Panel, 224, 227, 230
Image Sizing, 177
Info Overlay, 157
Inner Stroke, 230

K

Keyword List, 34
Keyword Set, 33
Keywording, 11, 33, 167

L

Layout, 54, 56-57, 205, 207-211, 216-217, 224-226, 228-232
Layout Style Panel, 231
Lens Corrections, 52, 124-129, 131-133, 135, 171, 196
Library Module, 4-7, 17, 33, 51, 87, 135, 138, 150-151, 156, 167, 176, 183, 193, 197-199, 201-203, 213-216, 248
Lights Out, 53, 167
Loupe, 5-7, 54, 56-57, 150, 156-157, 189, 194, 238, 240
Loupe Info, 156-157
Luminance, 89-94, 103, 120-122

M

Map Module, 200-203
Masking, 69, 71, 118-119, 121, 181, 246
Metadata, 9-10, 35-39, 44, 127, 131, 154, 163, 165, 171, 176, 178, 185, 190, 193, 202, 208, 213, 215, 220, 241-242

N

Navigator, 141, 168
Noise Reduction, 114-115, 117, 119-123, 161, 196

O

Output Sharpening, 178
Overlays, 54, 56-57, 217

P

Page, 56, 180, 184, 205, 207-210, 213, 224, 226-232
Page Panel, 209, 228-229, 231
Painter, 39
Photo Border, 230
Playback, 194, 198, 215, 219
Preferences, 106, 143, 148, 158-159, 161-165, 167, 169, 171-173, 194
Presets, 20, 22, 37, 94, 106, 125-126, 136-137, 139-143, 148, 159-160, 162, 172, 181, 184, 187
Print Adjustment, 225
Print Job Panel, 224, 230-231
Print Module, 223-225, 228, 233
Print Resolution, 224
Print Sharpening, 225
Profile Corrections, 127-129, 131
Publish, 171, 183-185, 199, 235
Publish Services, 183, 185, 199

Q

Quick Collection, 47, 153-154
Quick Develop, 7-8, 11, 19-20, 23, 29, 31, 77, 87, 195

R

Radial Filter, 65-66
Rating, 37, 39, 44, 153, 155-156, 215, 217
Red Eye, 60-61
RGB, 84-86, 163
Rotate, 13, 51-52, 54, 129, 133, 150, 155, 228-229

S

Saturation, 29, 64, 68, 73, 75, 89-94, 196, 223, 233
SD card, 190
Shadows, 17-18, 24, 26, 28, 68-70, 77, 80, 83, 85-87, 95-96, 99, 108-109, 116, 128, 183, 230, 233, 245, 248
Sharpening, 114-119, 121, 123, 178, 196, 225, 246
Slideshow Module, 36, 193, 197-198, 213-214, 216, 221
Smart Previews, 15, 239-240, 242
Snapshots, 136-137, 139-141, 143
Source, 2, 14, 44, 58-59, 171, 233
Split Toning, 86, 88-89, 91, 93, 95, 97, 99-101, 196
Spot Removal, 58-60, 248
Stacking, 30-31, 164
Standard Previews, 238
Survey View, 7

T

Tag, 35, 202-203
Target Collection, 42, 46-47, 153-154
Temperature, 17, 22, 51
Template Browser, 198, 214, 220-221, 223, 226
Tethering, 3, 186-187, 189, 191
Tint, 17, 21-22, 73, 95, 154, 251
Title, 3, 35-36, 44, 56, 156, 176, 178, 193, 214, 215, 219, 220, 246
Tone Control, 23-24
Tone Curve, 76-83, 85-87, 196, 245
Toolbar, 54-55, 137-138, 150, 167, 213-215, 220, 232
Tracklog, 203

V

Vibrance, 8, 17, 29, 89, 91, 193, 195-196
Video, 1-2, 10, 33, 157, 166, 169, 171, 177, 193-199, 213, 215, 219-221, 235
Vignetting, 96-99, 101, 127, 131, 133
Virtual Copies, 29-30, 41-42, 193, 197, 205, 214-215

W

Watermark, 171, 179-180, 211, 217, 228, 231
White Balance, 7-8, 17, 21-23, 89, 189, 195-196, 205

Whites, 17, 25, 27-28, 85, 106, 109, 195-196, 233, 247-248
Wi-Fi, 187, 189-191